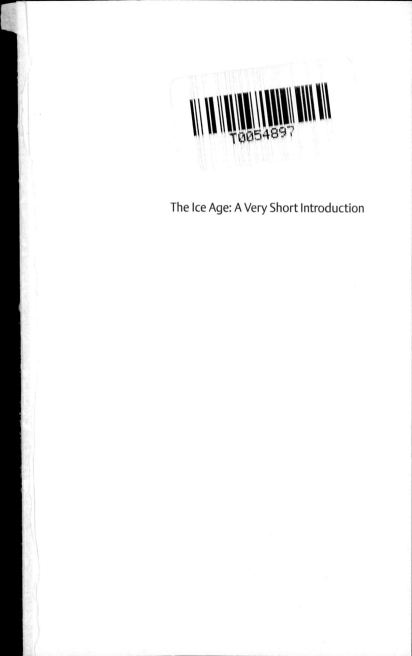

The Ice Age: A Very Short Introduction

VERY SHORT INTRODUCTIONS are for anyone wanting a stimulating and accessible way into a new subject. They are written by experts, and have been translated into more than 40 different languages.

The Series began in 1995, and now covers a wide variety of topics in every discipline. The VSI library now contains over 350 volumes—a Very Short Introduction to everything from Psychology and Philosophy of Science to American History and Relativity—and continues to grow in every subject area.

Very Short Introductions available now:

Available soon:

For more information visit our website
www.oup.com/vsi/

Jamie Woodward

THE ICE AGE

A Very Short Introduction

OXFORD
UNIVERSITY PRESS

OXFORD
UNIVERSITY PRESS

Great Clarendon Street, Oxford, OX2 6DP,
United Kingdom

Oxford University Press is a department of the University of Oxford.
It furthers the University's objective of excellence in research, scholarship,
and education by publishing worldwide. Oxford is a registered trade mark of
Oxford University Press in the UK and in certain other countries

Published in the United States of America by Oxford University Press
198 Madison Avenue, New York, NY 10016, United States of America

British Library Cataloguing in Publication Data
Data available

Library of Congress Cataloging in Publication Data
Data available

ISBN 978-0-19-958069-9

Printed and bound by
CPI Group (UK) Ltd, Croydon, CR0 4YY

Acknowledgements

I am very grateful to John Lewin and Mike Walker who read all the chapters and provided valuable suggestions and much encouragement. I must also thank Andrea Keegan and Emma Ma at OUP for keeping me on track. Several people (colleagues, friends, and family) read parts of the manuscript and provided helpful feedback—thanks to Jeff Blackford, Noel Castree, Colin Cook, Jason Dortch, Will Fletcher, Geraldine Shannon-Little, Danny Woodward, and Phillip Woodward. Nick Scarle drew all the line drawings with great skill and patience. I must also thank the following who helped enormously as I researched this book: Eliza Howlett and Keith Thomson at The Oxford Museum of Natural History who kindly showed me the Buckland archive and helped with the identification of items in the lithograph of 1823; Lucy Blaxland and Tony Simcock at the Old Ashmolean (now The Museum of the History of Science in Oxford) who showed me the room where Buckland lectured; The John Rylands Library on Deansgate, Manchester, where I viewed Agassiz's *Études sur les glaciers*; Patrick Boylan, for valuable discussions on Buckland and Agassiz; and Mike Hall for many enjoyable conversations about his career with Nick Shackleton. Finally, I must thank my wonderful family for not only giving me the space to complete this book, but for commenting on each chapter (Jenny), and for sitting on erratic boulders (Sam and Alex). They are still wondering why this very short introduction took such a very long time.

This book is dedicated to all the staff and pupils of Woolston High School (1957–2012) in Warrington. Its Geography Department and Outdoor Pursuits Club—both led by Paul Layfield in my era—sparked my interest in landscapes and environmental change.

Contents

List of illustrations

calibration - ODP 677. IGBP PAGES/ World Data Center-A for Paleoclimatology Data Contribution Series # 96-018. NOAA/NGDC Paleoclimatology Program, Boulder, CO, USA. b) modified from Shackleton, N.J. and Opdyke, N.D. (1973) Oxygen isotope and palaeomagnetic stratigraphy of equatorial Pacific core V28-238: oxygen isotope temperatures and ice volume on a 10^5 and 10^6 year scale. Quaternary Research, 3, 39–55. c) modified from Roucoux, K.H., Shackleton, N.J., de Abreu, L., Schönfeld, J., and Tzedakis, P.C. (2001) Combined marine proxy and pollen analyses reveal rapid Iberian vegetation response to North Atlantic millennial-scale climate oscillations. Quaternary Research, 56, 128–132

The publisher and author apologize for any errors or omissions in the above list. If contacted they will be happy to rectify these at the earliest opportunity.

It is now well ascertained that during a comparatively recent geological period, the climate of the northern hemisphere was much colder than at present, and that in the British Islands, as well as in other countries where glaciers are now unknown, the land was enveloped in snow and ice. This part of the geological record is known as the Ice Age or Glacial Period.

Archibald Geikie (1887), p. 242

Prologue

A letter from Darwin

When Adam Sedgwick died in January 1873 at the age of 87 he had held the Woodwardian Professorship of Geology at Cambridge for 55 years. Sedgwick was one of the founding fathers of the discipline. One of the nine candidates to succeed him, William Boyd Dawkins (1837–1929), a lecturer in geology in Manchester and an expert on the fossil remains of ice age mammals, asked Charles Darwin to support his application. Darwin responded as follows:

> DOWN, BECKENHAM, KENT
> February 3rd 1873
> MY DEAR SIR,
> I have great pleasure in expressing my opinion that you are very well fitted to fill the Woodwardian Chair in Cambridge, now vacant by the death of its former venerated occupier. You have paid close attention to the geological history of the more recent periods, and I think everyone will admit that these present an extraordinary amount of difficulty; so that your success in this line of research offers an excellent test of your ability. It will also, I think, be admitted that the study of the more recent periods is not only very difficult, but of the highest importance. Therefore I earnestly hope that you may be successful in your application, and if so, I do not

doubt that you will be the means of encouraging the study of
geology in the University.

Believe me, my dear Sir,

Yours sincerely,

CH. DARWIN

Boyd Dawkins had been appointed in Manchester in 1869 on the
recommendation of Thomas Huxley (1825–95). His reputation
was built on the study of fossil fauna from the Quaternary
Period—the most recent and shortest of the geological periods
that is synonymous with the environmental fluctuations of the
Great Ice Age. This period, the last 2.58 million years of Earth
history, is the subject of this book.

It was the discovery of fossil animal bones of cold-adapted species
such as reindeer, muskox, and arctic fox—which now live in the
treeless landscapes of the Arctic tundra—that provided some of
the earliest clues that the climate of the now temperate zone had
been much colder in the recent geological past. But for much of
the 19th century climate change was a controversial topic and
most of the leading geologists were climate change sceptics.
Darwin's own scientific career ran alongside, and from time to
time contributed to, the great 19th-century debate on whether
glaciers had been more extensive and climate much colder in the
recent geological past. The glacial theory, as it became known,
produced one of the most heated and protracted scientific
controversies of the 19th century. It was an exciting time of
colourful characters and big ideas.

This is not a book about Darwin, but, as we shall see, he was
uniquely qualified to comment in the 1870s on the significance of
ice age research and on the suitability of William Boyd Dawkins to
follow in the footsteps of Adam Sedgwick. His brief letter of
February 1873 and the background to its assertions are therefore
worthy of some elaboration here. They illuminate key aspects of the
glacial debate, show why ice age research is a critical part of the

natural sciences, and provide valuable context for much of
this book.

It is important to remember that Darwin regarded himself as a
geologist when he boarded the *Beagle* in December 1831 and that he
returned to England in 1836 the year before the Swiss naturalist
Louis Agassiz (1807–73) presented his bold idea of a great ice sheet
covering much of Europe, Siberia, and North America. Agassiz
toured Scotland in the autumn of 1840 with the Oxford geologist
William Buckland (1784–1856) seeking evidence for the former
action of glaciers. Both presented papers on their glacial research to
the Geological Society in London later that year. Because Darwin
was secretary of the Geological Society from 1838 to 1841, he was
fully immersed in the machinations of the glacial debate in Britain.

By pointing out that the study of the most recent periods is 'of the
highest importance' Darwin provides a powerful endorsement for
Boyd Dawkins' research agenda and for the study of the glacial
period more generally—this is what we now call Quaternary
science. In the 1870s many scholars were still coming to terms
with the notion that humans had lived alongside now extinct ice
age beasts such as the mammoth and woolly rhino. Darwin, more
than anyone in the 1870s, was aware of the enormous importance
of developing an improved understanding of the recent geological
past, its environmental fluctuations, and the antiquity of humans
and other animal species. Even though he was frustrated by the
sparseness of the human fossil record, Darwin was also convinced
that the environmental changes of the recent geological past
formed the backdrop to the evolution of our own species, even if
the precise timescale involved was yet to be established. Exploring
the relationship between climate change and human evolution
remains a key field of enquiry today.

In 1839, following geological fieldwork in the Scottish Highlands
the previous summer, Darwin published a paper on the enigmatic
landscape features near Ben Nevis known as the Parallel Roads of

1. The Parallel Roads of Glen Roy

Glen Roy (Figure 1). Several theories had been advanced to explain the origin of the distinctive ledges that run along both sides of the valley. Darwin argued that Glen Roy was once an inlet of an ancient sea that covered much of central Scotland. He suggested that the Parallel Roads were shorelines marking changes in the level of this sea.

Barely a year after its publication, however, and to his great dismay, Darwin's interpretation was challenged. Agassiz, who had seen similar features in the Alps, argued that the valley had been blocked by a large glacier during the ice age and the shorelines marked the former levels of an impounded glacial lake. Agassiz's glacial model prevailed. Darwin only abandoned his marine theory, with great reluctance, in the 1860s. He later called it 'one long gigantic blunder'.

In August 1831, to improve his knowledge of geology, Darwin had accompanied Sedgwick on the great professor's regular summer fieldwork stint in North Wales. Darwin was 22 and had just graduated from Cambridge. The two of them spent several days in the mountains of Snowdonia in the heart of what we now know is a classic glacial landscape. But Sedgwick and Darwin only had eyes for the very old solid geology of the Cambrian Period—the earliest known geological period at that time, as defined and named by Sedgwick himself. The two geologists, master and pupil, were surrounded by the work of ancient glaciers—yet both were oblivious to all of it. Writing in his diary many years later (1876) Darwin reflected on his own glacial epiphany in Snowdonia:

> We spent many hours in Cwm Idwal, examining all the rocks with extreme care, as Sedgwick was anxious to find fossils in them; but neither of us saw a trace of the wonderful glacial phenomena all around us; we did not notice the plainly scored rocks, the perched boulders, and the lateral and terminal moraines. Yet these phenomena are so conspicuous that, as I declared in a paper published many years afterwards in the 'Philosophical Magazine' (1842), a house burnt down by fire did not tell its story more plainly than did this valley. If it had still been filled by a glacier, the phenomena would have been less distinct than they now are.

Not only did Darwin regard the study of 'the more recent periods' as being 'of the highest importance', his comment about the 'extraordinary amount of difficulty' involved in studying the most recent deposits and landforms in Earth history is especially poignant given his own profound disappointment over Glen Roy and his reflections on a missed opportunity at Cwm Idwal. This was a truly formative period for Darwin. The events of 1839 to 1842 left a lasting impression as he witnessed the power of a bold new theory.

Boyd Dawkins did not get the chair in Cambridge. He stayed in Manchester for the rest of his career pursuing many interests as the Curator of the Manchester Museum and the first Professor of Geology in the University. He was knighted in 1919. Boyd Dawkins held some controversial views on the role of glacial ice in Britain, but he is perhaps best remembered for his pioneering work on the fossil remains of ice age animals from British limestone caves as part of the golden age of Victorian geology. He also made important contributions to Palaeolithic archaeology. It was during his career that the foundations were laid for the modern interdisciplinary study of Quaternary ice age environments that we know today.

This book then is concerned with the remarkable environmental changes that have taken place during the Great Ice Age of the Quaternary Period. It explores the evolution of ideas and changing approaches to the study of the recent geological past; from the pioneers of the 19th century who first recognized that glaciers had once been much more extensive, to the pioneers of the 20th century who transformed our understanding of global climate change by extracting exquisitely detailed records from ocean sediments and ice cores. Most of all it aims to show why the study of the ice age is exciting and rewarding and is still 'of the highest importance' 140 years after Darwin wrote that letter.

Chapter 1
The Quaternary ice age

Nothing excites the imagination more than the study of the Quaternary.

Maurice Gignoux (1955)

Early ideas: elephants or mammoths?

In 1727, the physician, compulsive collector of curiosities, and shrewd investor in London property, Sir Hans Sloane (1660–1753), published *An Account of Elephants' Teeth and Bones Found under Ground*. It reported one of the first systematic investigations of mammoth fossils—describing finds of large bones and tusks (of both elephants and mammoths) from Quaternary deposits in Britain, France, and Siberia. Sloane's paper also included items from his own *Catalogue of Quadrupeds*. In the same year, Sloane succeeded Sir Isaac Newton as president of the Royal Society. He was an enormously influential figure and his work was widely read.

The discovery of elephant-like beasts in the superficial deposits of Europe and Siberia was rather problematic for 18th-century scholars because such creatures only lived in the tropics and climate change was not on the agenda. Sloane and his contemporaries believed that these animals had perished in the Biblical deluge and their carcasses had been washed northwards by the flood. In 1728, the German zoologist Johann Philipp Breyne (1680–1764) set out this position very clearly as the only reasonable explanation for the presence of mammoth bones in the frozen floodplains of Siberia:

> That those Teeth and Bones of *Elephants* were brought thither by no other Means but those of a Deluge, by Waves and Winds, and left behind after the Waters return'd into their Reservoirs, and were buried in the Earth, even near to the Tops of high Mountains. And because we know nothing of any particular extraordinary Deluge in those Countries, but of the universal Deluge of *Noah*, which we find described by Moses; I think it more than probable, that we ought to refer this strange Phenomenon to the said Deluge. In such Manner, not only the holy Scripture may serve to prove natural History; but the Truth of the Scripture, which says that *Noah's* Flood was universal, a thing which is doubted by many, may be proved again by natural History.

Such certainty (and circularity) was typical of the period and these beliefs were strengthened by the fate of an exotic gift to the Tsar of Russia. In 1713, Peter the Great (1672–1725) was presented with an Asian elephant by the Persian Ambassador. Unfortunately the poor creature soon perished in the brutal St Petersburg winter and ended up stuffed in the Imperial Museum. Elephants just weren't cut out for the frozen north.

In 1753, following an act of Parliament, Sloane's extraordinary collection of curiosities—mammoth tusks and all—was opened to the public alongside King George II's library as the British Museum in Bloomsbury. For the rest of the century, mammoth bones and teeth continued to be mysteries of an antediluvian world. It took an astonishing discovery from the frozen wastelands of Siberia (that would itself become a ground-breaking museum exhibit) to put such curiosities on a more secure scientific footing.

Cuvier's insights and the Adams Mammoth

In 1799, just weeks apart, two quite remarkable discoveries were made in the lower reaches of large rivers that would, in very different ways, transform our understanding of the past. In the same summer that the Rosetta Stone was unearthed from the mud of the Nile Delta, a hunting party in eastern Siberia discovered a frozen mammoth carcass in a steep bank of the mighty River Lena not far from the Arctic Ocean. As the river sediments thawed the ancient beast was slowly exhumed—within a few years the entire carcass had slumped to the ground at the foot of the bank. Preyed upon by polar bears and wolves, the decomposing mass was partly defleshed and some of the large bones were dragged away. In 1804, the leader of the hunting party, Ossip Shumakhov, returned to the site, sawed off the giant tusks, and sold them to an ivory merchant for 60 roubles. News of the extraordinary discovery soon reached St Petersburg.

In the summer of 1806, Mikhail Adams (1780–1838), a Russian botanist under the patronage of Tsar Alexander I, trekked to the site to recover what was left of the mammoth. Although the trunk was missing, the skeleton was almost complete and large sections of skin with its woolly fleece and long hair were still intact. Many parts of the skeleton were firmly held together by cartilage, ligaments, and skin. Over 16.5 kg of mammoth wool and hair were gathered up from the carcass and the sandy river margin. Adams claimed he managed to buy back the original 3-metre-long tusks. He then shipped everything to St Petersburg—a distance of some 11,800 km. The Adams Mammoth is still one of the most complete mammoth carcasses ever found. Its skin and thick hairy fleece were exceptionally well preserved and samples were sent to museums across Europe.

Barely a decade earlier, in 1796, the brilliant French anatomist Georges Cuvier (1769–1832) became the first person to argue that the mammoth was a distinct species quite different from modern elephants. He was therefore able to declare that this species was extinct. Cuvier established extinctions as a scientific fact. To explain the extinctions he observed in the fossil record he became the most prominent advocate of what later became known as catastrophism, suggesting that new species were created after sudden events wiped out earlier ones. The mammoth was one of the first extinct animals to be discovered and investigated scientifically—the first dinosaur was not named until the 1820s. Cuvier is the father of modern vertebrate palaeontology and comparative anatomy—it was said he could recreate the anatomy of an extinct creature from just a handful of bones.

Most importantly, the well preserved hair and wool remains of the Adams Mammoth convinced Cuvier that, unlike Peter the Great's elephant, this large, fleshy creature was adapted to the frozen world of the Arctic. It had lived and died in the frozen north—there was no need to invoke post-mortem carriage from tropical climes by Biblical floods. These seminal conclusions placed the mammoth as a key fossil indicator of Arctic conditions. Discoveries of mammoth bones in the Quaternary deposits of the temperate latitudes now assumed much greater significance. Cuvier's conclusions laid down fundamental principles for what would become ice age research. The modern era of debating climate change had begun.

In 1808, the Adams Mammoth skeleton was reassembled by the German naturalist and explorer Wilhelm Tilesius von Tilenau (1769–1857). Tilesius used an elephant skeleton in the St Petersburg museum collection to help him rebuild the mammoth, replacing two missing leg bones with carved wooden replicas. Imagine the frustration on finding that the first mammoth kit had pieces missing! Standing about 3 m in height and over 5 m long, it was the skeleton of an adult male that died when it was about 45 years old. The magnificent reconstruction was put on display in the Imperial Museum in St Petersburg where it towered over the stuffed carcass of Peter the Great's Asian elephant that stood nearby (Figure 2).

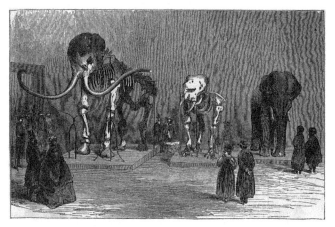

2. The Adams Mammoth

Tilesius made a detailed drawing of the skeleton that was widely distributed in both scholarly publications and the popular press. The Adams Mammoth began to feature prominently in university lectures in Europe and North America. This was one of the earliest attempts to reconstruct the skeleton of an extinct animal and the first time that a mammoth skeleton had been put on public display. Unfortunately, as the image of the museum gallery clearly shows, the enormous mammoth tusks were mounted pointing outwards instead of inwards. This mistake was not rectified until 1899—exactly one hundred years after the carcass was first discovered protruding from the frozen banks of the Lena River. Despite this error, between them, Cuvier, Shumakhov, Tsar Alexander I, Adams, and Tilesius had paved the way for the mammoth to become *the* iconic symbol of the Quaternary ice age.

The nature of the Quaternary

What the Quaternary Period lacks in length is more than compensated by the wonderful variety and exquisite detail of its sedimentary records. These records, on land and on the ocean floor, preserve evidence of profound and global-scale changes in climate, landscapes, and ecosystems. It is important to appreciate from the outset that the Quaternary ice age was not one long episode of unremitting cold climate. The Quaternary is all about change. How much, how often, how fast?

By exploring the landforms, sediments, and fossils of the Quaternary Period we can identify *glacials*: periods of severe cold climate when great ice sheets formed in the high middle latitudes of the northern hemisphere and glaciers and ice caps advanced in mountain regions around the world. We can also recognize periods of warm climate known as *interglacials* when mean air temperatures in the middle latitudes were comparable to, and sometimes higher than, those of the present. As the climate shifted from glacial to interglacial mode, the large ice sheets of Eurasia and North America retreated allowing forest biomes to re-colonize the ice free landscapes.

It is also important to recognize that the ice age isn't just about advancing and retreating ice sheets. Major environmental changes also took place in the Mediterranean region and in the tropics. The Sahara, for example, became drier, cooler, and dustier during glacial periods yet early in the present interglacial it was a mosaic of lakes and oases with tracts of lush vegetation. A defining feature of the Quaternary Period is the *repeated* fluctuation in climate as conditions shifted from glacial to interglacial, and back again, during the course of the last 2.5 million years or so. A key question in ice age research is why does the Earth's climate system shift so dramatically and so frequently?

The great ice sheets

Today we have large ice masses in the Polar Regions, but a defining feature of the Quaternary is the build-up and decay of continental-scale ice sheets in the high middle latitudes of the northern hemisphere. Figure 3 shows the Laurentide and Cordilleran ice sheets that covered most of Canada and large tracts of the northern USA during glacial stages. Around 22,000 years ago, when the Laurentide ice sheet reached its maximum extent during the most recent glacial stage, it was considerably larger in both surface area and volume (34.8 million km^3) than the present-day East and West Antarctic ice sheets combined (27 million km^3). With a major ice dome centred on Hudson Bay greater than 4 km thick, it formed the largest body of ice on Earth. This great mass of ice depressed the crust beneath its bed by many hundreds of metres. Now shed of this burden, the crust is still slowly recovering today at rates of up to 1 cm per year. Glacial ice extended out beyond the 38th parallel across the lowland regions of North America. Chicago, Boston, and New York all lie on thick glacial deposits left by the Laurentide ice sheet.

From an archaeological perspective, understanding the changing geography of the globe during glacial and interglacial stages is of the utmost importance. With huge volumes of water locked up in the ice sheets, global sea level was about 120 m lower than present at the Last Glacial Maximum (LGM), exposing large expanses of continental shelf and creating land bridges that allowed humans, animals, and plants to move between continents. Migration from eastern Russia to Alaska, for example, was possible via the Bering land bridge.

Understanding the shifting dimensions of the Cordilleran ice sheet and the Pacific coastal zone (and the timing of these changes) is critical in this respect and especially during the last deglaciation as ice sheets retreated and sea levels began to rise. Did humans move south along the west coast of Canada or did they make use of an 'ice free corridor' between the Laurentide and Cordilleran ice sheets? These are the two most likely routes for the migration of humans into North America at the end of the last glacial period (Figure 3).

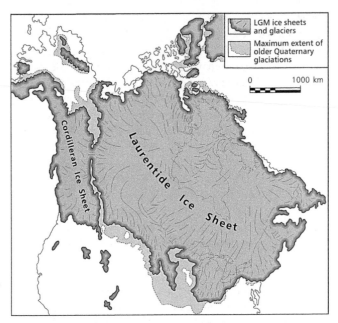

3. North American ice sheets at the Last Glacial Maximum

Box 1
The cryosphere

Earth is unique in our Solar System because water is abundant in all three phases—in solid, liquid, and gaseous form—and is continuously cycled around the planet. Without large oceans of liquid water and a fully active hydrological cycle, it would not be possible to build ice sheets and melt them again. The frozen parts of the Earth system are known as the cryosphere: from the Greek *cryos* meaning cold. Today, freshwater accounts for barely 2.5 per cent of the water on Earth (the other 97.5 per cent is salty ocean water) and almost 70 per cent of this freshwater is bound up as ice sheets and permanent snow. The ice sheets of Antarctica and

Greenland cover an area of about 15.7 million km² and have a combined volume of almost 30 million km³. If all this ice became liquid, sea level would rise more than 65 m. We are still very much within a glacial epoch.

Ice age Europe

Large ice sheets also developed in Europe. Figure 4 shows transects from the high Arctic to the Mediterranean Sea under interglacial and glacial conditions. Much of the continent is covered in woodland during interglacial stages (note that this shows vegetation without any disturbance by humans) and sea levels were similar to or higher than the present. At the LGM, however, European geography was transformed. The Baltic and North Seas were dry land and Britain was connected to mainland Europe. Beyond the British and Scandinavian ice sheets, much of central and northern Europe was a treeless tundra steppe habitat. As in North America, large herbivores like mammoth, reindeer, and bison roamed the landscapes to the south of the ice sheets and were tracked by groups of Palaeolithic hunters. Such reconstructions of Quaternary geography illustrate how landscape processes and ecosystems across the northern hemisphere were completely reorganized as the climate system shifted between glacial and interglacial modes.

The British Isles lie in an especially sensitive location on the Atlantic fringe of Europe between latitudes 50 and 60° north. Because of this geography, the Quaternary deposits of Britain record especially dramatic shifts in environmental conditions. The most extensive glaciation saw ice sheets extend as far south as the Thames Valley with wide braided rivers charged with meltwater and sediment from the ice margin. Beyond the glacial ice much of southern Britain would have been a treeless, tundra steppe environment with tracts of permanently frozen ground (Figure 4).

The Ice Age

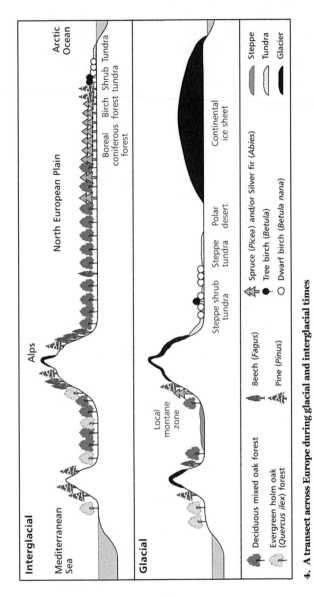

4. A transect across Europe during glacial and interglacial times

During warm interglacial stages these landscapes would have been transformed. Broad-leaved deciduous woodland with grassland was the dominant vegetation at these times. In the warmest parts of interglacials thermophilous (warm-loving) insects from the Mediterranean were common in Britain whilst the large mammal fauna of the Last Interglacial (c.130,000 to 115,000 years ago) included even more exotic species such as the short tusked elephant, rhinoceros, and hippopotamus. In some interglacials, the rivers of southern Britain contained molluscs that now live in the Nile Valley. For much of the Quaternary, however, climate would have been in an intermediate state (either warming or cooling) between these glacial and interglacial extremes. Exploring how ecosystems responded to these climate changes is a fundamental objective of Quaternary research.

The transects in Figure 4 are highly simplified but they provide an important statement on the wholesale and continental-scale reorganizations of life and landscapes that we have to conceptualize in the Quaternary Period. They show the broad pattern of vegetation change, but think also about how insects, birds, mammals, and fish, not to mention humans, coped with such large-scale reconfigurations of their ecosystems. Later in this book we will tackle some fundamental questions: How many times have these grand reorganizations taken place? What was the pace of change? Can we identify a typical state for planet Earth in the Quaternary?

Other important questions follow from such reconstructions. How can we reconstruct the dimensions of an ice sheet that no longer exists? How many glacial periods were there during the Quaternary and how long did they last? How can Quaternary records help us to better understand recent climate trends? This book will explore these questions and consider the kinds of information we need to tackle them.

Evidence of glacial action in the landscape

Glaciers produce a distinctive assemblage of erosional and depositional landforms (Figure 5). The erosional features are best preserved in hard rocks such as granites and limestones. In upland landscapes the tell-tale signs of the former presence of glacial ice include bowl-shaped, glacially eroded bedrock hollows known as cirques (also known as cwms or corries). Cirques are commonly backed by steep bedrock cliffs. An arête is a sharp ridge that forms between neighbouring cirques. During ice free periods cirques are commonly occupied by a lake (tarn) enclosed by a moraine or bedrock lip. Figure 5 shows a number of cirques and hanging valleys along a deep U-shaped valley. The latter was

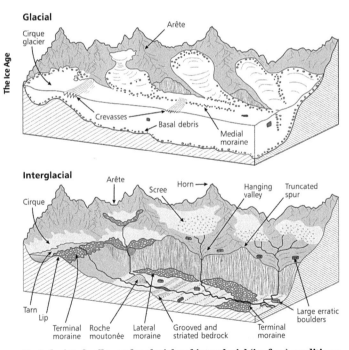

5. **A glaciated valley under glacial and interglacial (ice free) conditions**

occupied by the main glacier. Polished and scratched bedrock surfaces are found in all parts of this upland landscape.

Landforms produced by the deposition of glacial sediment include lateral and terminal moraines, drumlins, and large erratic boulders. Darwin observed many of these features in Cwm Idwal in North Wales in 1841. Thick sheets of boulder clay (also known as till) cover large tracts of the landscape that were once covered by Quaternary ice sheets. Glacial deposits recognized in ancient hard rocks much older than the Quaternary are known as tillites. Many of the terms for glacial features were first coined in the Alps by German and French speaking observers. Meltwater streams draining from glaciers also produce distinctive fluvial sediments and landforms.

Glaciologists make a distinction between three main types of glacier (valley glaciers, ice caps, and ice sheets) on the basis of scale and topographic setting. A glacier is normally constrained by the surrounding topography such as a valley and has a clearly defined source area. An ice cap builds up as a dome-like form on a high plateau or mountain peak and may feed several outlet glaciers to valleys below. Ice sheets notionally exceed 50,000 km^2 and are not constrained by topography. In Antarctica most ice is transported to the coast via a series of fast moving ice streams. Large blocks of ice can calve off into the ocean to form icebergs.

The Quaternary in the geological timescale

The position of the Quaternary Period in the geological timescale is shown in Figure 6. It sits at the top of the pile. The final period of the Cenozoic Era, it accounts for just a tiny fraction of Earth history—about 0.056 per cent of the 4.6 billion years since the formation of our planet. The term Quaternary was first introduced in the 18th century by the Italian geologist Giovanni Arduino (1714–95). It is the only survivor from the time when the geological record was divided into four parts: Primary, Secondary, Tertiary, and Quaternary.

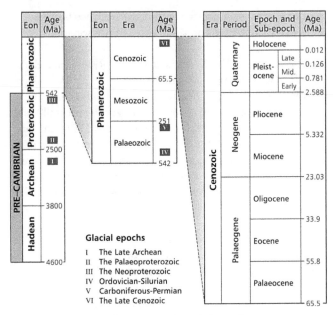

Eon	Age (Ma)	Eon	Era	Age (Ma)	Era	Period	Epoch and Sub-epoch	Age (Ma)

6. The Quaternary Period in the geological timescale

The Quaternary is sub-divided into the Pleistocene and Holocene epochs (Figure 6). The end of the Pleistocene marks the end of the last glacial period and the beginning of the Holocene (11,700 years ago)—the current interglacial—when the large ice sheets of Europe and North America had melted away for the last time. This was the last time the Earth's climate system shifted from glacial to interglacial conditions. The Holocene is often referred to as the post-glacial and includes the present day.

Glacial epochs in Earth history

We live in unusual times. For more than 90 per cent of its 4.6-billion-year history, Earth has been too warm—even at the poles—for ice sheets to form. Ice ages are not the norm for our planet. Periods of sustained (over several million years) large-scale glaciation can be called glacial epochs. Tillites in the geological record tells us that the Quaternary ice age is just one of at least six great glacial epochs that have taken place over the last three billion years or so (Figure 6). The Quaternary itself is the culmination of a much longer glacial epoch that began around 35 million years ago (Ma) when glaciers and ice sheets first formed in Antarctica. This is known as the Cenozoic glacial epoch.

There is still much to learn about these ancient glacial epochs, especially the so-called Snowball Earth states of the Precambrian (before 542 Ma) when the boundary conditions for the global climate system were so different to those of today. Exploring when and why large-scale glaciation was initiated at much earlier times in our planet's history is also important because it can shed light on the nature of climates for periods which don't have fossil records.

Quaternary sediments and fossils

This book is concerned with the Quaternary ice age—it has the richest and most varied records of environmental change. Because its sediments are so recent they have not been subjected to millions of years of erosion or deep burial and metamorphism. Most Quaternary deposits are therefore not lithified (turned to hard rocks): they are soft sediments that can be sampled at high resolution with a knife rather than a hammer. They are often rich in fossils, large and small, that have undergone little or no alteration since burial. In aquatic settings, such as lakes and peat bogs, organic materials such as insects, leaves, and seeds, as well as microfossils such as pollen and fungal spores can be exceptionally well preserved in the sediment record. This allows us to create very detailed pictures of past ecosystems under glacial and interglacial conditions. This field of research is known as Quaternary palaeoecology.

Early observers of Alpine landscapes recognized the work of glaciers in places that were then ice free. They made the link between the processes, sediments, and landforms they observed at the margins of active glaciers and landscape features they saw further down valley—often great distances from the snouts of the nearest modern glaciers. But postulating longer and thicker glaciers in the Alps was one thing, making the case for Quaternary glaciers in the mountains of the British Isles, and for immense sheets of ice covering much of Europe and North America was, as we shall see, quite another.

Chapter 2

Erratic boulders and the diluvium

Time it does not matter
But time is all we have
To think about
                    Deep Purple (2013)

This chapter explores some of the key ideas in geology in the early 19th century. It sets out some of the problems faced by early geologists as they attempted to make sense of the landscapes around them and the most recent part of the geological record. Explaining the origin of the erratic boulders and superficial deposits that we now know date to the Quaternary Period proved to be especially problematic. An understanding of this background is necessary in order to appreciate the formidable obstacles that confronted those who championed the glacial theory in Europe and North America.

Box 2
What is an erratic boulder?

An erratic (from the Latin *errare*: to wander or go astray) is a piece of rock that has been transported from its place of origin. Erratic boulders may sit on the modern land surface or be deeply buried within thick deposits of boulder clay. Many erratics stand out because they lie on bedrock that is very different to their source. Figure 7 shows a wonderful example of an erratic boulder of brown Silurian sandstone sitting on a pedestal of pale grey Carboniferous limestone in the Yorkshire Dales of northern England. The boulder is foreign and immediately raises three questions: Where did it come from, and how and when was it transported? Geologists spend a lot of time thinking about provenance, processes, and time. Erratics are normally associated with transport by glaciers or ice sheets, but in the early 19th century mechanisms such as the great deluge or rafting on icebergs were commonly invoked. It has been argued that the famous bluestones of Stonehenge are glacial erratics. A lump of Dedham Granodiorite, better known as Plymouth Rock, the iconic symbol marking the arrival of the Pilgrim Fathers in New England in 1620, is also a glacial erratic.

As a scientific discipline, geology was still very much in its infancy in the latter part of the 18th century. Collectors still hoarded fossils and geological curiosities of all kinds, much as they had done in previous centuries, yet the purpose and practice of

7. An erratic boulder in northern England

geological research and concepts of time and process were still only very loosely defined. For the vast majority in the Christian world the Bible provided all that one needed to know about the origin of the Earth and the creatures upon it. By tracing the progress of geology in the early 19th century we can see the uneasy passage from medieval to modern ways of thinking about the history of the Earth. As far as glaciers were concerned, these 'rivers of ice' were viewed as curiosities of the remote and little explored Polar and high mountain zones. They were certainly not considered to have played a role in landscape formation or sediment transport *beyond* these areas.

By the turn of the 18th century, as theories of the Earth were being fashioned and debated, a more systematic approach to the field and laboratory study of rocks and fossils was emerging. William 'Strata' Smith (1769–1839), for example, was developing his field mapping strategy that led to the publication of the first geological map of Britain in 1815. The growing influence of Georges Cuvier's work on comparative anatomy was transforming fossil hunting into scientific vertebrate palaeontology. Learned societies devoted to geological matters were set up and natural history was gaining a foothold in universities in Europe and North America, but hardly anyone was thinking about ice sheets or glaciers.

Hutton's *Theory of the Earth*

A new theoretical framework for geology came out of the Scottish Enlightenment when the Edinburgh-born physician, farmer, and self-taught geologist, James Hutton published his *Theory of the Earth* in 1795. Hutton was part of a remarkably creative circle of Edinburgh intellectuals that included David Hume, Adam Smith, John Playfair, and the poet Robert Burns. All were members of the famous Oyster Club that met regularly across the city to discuss the latest ideas in science and philosophy. Joseph Black (1728–99), the chemist and discoverer of carbon dioxide (CO_2), was also a member of this illustrious group. Black is often called

Hutton's 'chemistry teacher' and in later chapters we shall explore the fundamental role played by the gas he discovered in the climate changes of the Quaternary ice age. James Hutton changed the way we think about landscape evolution and his philosophy was to impact directly upon the debate over the glacial theory. He also put forward some important early ideas on glaciation that have not been widely recognized.

To explain the origin of landscapes and the geological record, Hutton advocated a dominant role for 'existing causes'—in other words: processes that were known and could be observed in the field. He had no time for groundless and fantastic explanations such as enormous flood waves or violent convulsions of the Earth's crust. Hutton argued that weathering, erosion, and sedimentation operated only very slowly over exceedingly long timescales to form the landscapes of the Earth's surface. Change took place in small increments but, with enough time, the cumulative impact of these small changes could build great mountain ranges and then erode them away. Thus, by advocating *uniformity* in the rate at which processes operated, Hutton rejected change by abrupt catastrophic phenomena that were beyond human experience. The idea that our knowledge of the present should always guide the way we interpret the past was taken up and championed most famously by Charles Lyell in the 1830s. This *uniformitarian* approach was seen as the antidote to *catastrophism* and it became the dominant paradigm in 19th-century geology. We shall look at some of Lyell's ideas a little later.

Hutton famously stated that the relentless recycling of rocks and sediments took place with 'no vestige of a beginning, no prospect of an end'. This implied of course that the Earth must be very ancient indeed. Thus Hutton introduced the idea of *deep time*, believing that the formation of geological strata, life forms, and landscapes required immensely long spans of time and the inexorable cycling of rocks as mountains were worn down and rivers carried their loads to the sea.

Hutton on glaciation

In his *Theory of the Earth* (1795), Hutton commented on the erosive and transporting powers of the glaciers of the Alps—the first British geologist to do so. He argued that the very steep slopes of the highest mountains indicated that the sculpting action of glaciers may have been more active in the past at a time when:

> There would then have been immense valleys of ice sliding down in all directions towards the lower country, and carrying large blocks of granite to a great distance, where they would be variously deposited, and many of them remain an object of admiration to after ages, conjecturing from whence, or how they came. Such are the great blocks of granite which now repose upon the hills of Salève.

Salève (1,379 m) is a flat-topped limestone landscape close to the French-Swiss border overlooking Geneva. Geologically it is part of the Jura chain. Hutton understood that glaciers were powerful agents of erosion and sediment transport. By stating that the great granite blocks strewn across the flanks of the Jura Mountain limestones were conveyed there by ice, he was clearly stating that the glaciers of the Alps had formerly been much more extensive. Unfortunately, this tantalizing glimpse of a latent glacial theory was not developed further in his lifetime.

Hutton cited extensively from the works of Horace-Bénédict de Saussure (1740–99), the great Alpine explorer and scientist, but his early ideas on glaciation have often been overlooked. Hutton was the first person writing in English to put forward the idea of a more extensive Alpine glaciation. These insights were advanced long before Agassiz proposed his own grand theory.

The problem of erratic boulders

Enormous erratic boulders, including those of the Jura described by Hutton and others, were well known to 18th- and 19th-century geologists. Some were the size of a three-storey house. Huge erratics of Canadian provenance were well known across the northern states of the USA and colossal boulders of Scandinavian origin were recorded along the coastal lowlands from Estonia to the Netherlands. How had these massive hunks of rock traversed the Baltic and North Seas? Their origin was a source of lively and protracted debate on both continents. The progress of this debate was central to the emergence of the glacial theory.

Early observers of Alpine glaciers had noted the presence of large boulders on the surface of active glaciers or forming part of the debris pile at the glacier snout. These were readily explainable, but erratic boulders had long been noted in locations that defied rational explanation. The erratics found at elevations far *above* their known sources, and in places such as Britain where glaciers were absent, were especially problematic for early students of landscape history. Explaining the origin of these far-travelled blocks presented problems for both catastrophists and uniformitarians.

Figure 8 shows a colossal slab of granite weighing several thousand tons lying on the limestone hillside that slopes gently down from Monte San Primo (1,682 m) to Lake Como in northern Italy. This sketch shows just one of the many enormous erratics mapped by

8. An erratic block on the southern flank of the Alps near Lake Como

the pioneering British geologist Henry de la Beche (1796–1855) around Como and Bellagio during fieldwork in the region in 1829.

These blocks originated far to the north in the granite core of the central Alps. In his *Geological Manual* (1832) De la Beche pointed out that many theories had been advanced to explain the origin and emplacement of such erratics, but none had yet proved satisfactory. He then advanced a theory of his own involving transport in a great flood charged with floating masses of ice:

> It has been above remarked, that the Alpine erratic blocks frequently occur in groups. To present a general explanation of this phenomenon would, at present, be somewhat difficult; but it may be asked, as a mere conjecture, whether masses of floating ice charged with blocks and other detritus, rushing down to the great valleys into the more open country of lower Switzerland, might not be whirled about by the eddies, and the icy masses be destroyed by collision against each other, so that groups of blocks would afterwards be found beneath the places where the whirlpools had existed.

A huge deluge (with or without floating bergs of ice) was commonly invoked to explain the disposition of such boulders and many saw them as more hard evidence in support of the Biblical flood. A flood wave of quite extraordinary power would be needed to move such enormous slabs of rock. In his *Principles of Geology*, Charles Lyell proposed a similar mechanism to that of De la Beche involving boulder-bearing icebergs released from alpine lakes in great floods triggered by the breaching of natural barriers. Such explanations for the erratic boulders found in the uplands and on the flanks of mountain ranges such as the Jura are typical of the period, but one can sense the frustration of De la Beche at the inadequacy of existing 'conjectures' (including his own) and it is worth briefly exploring the reasons why.

De la Beche noted the angular character of the block of granite shown in Figure 8 and observed that if it had been moved in a large flood, the edges and corners would have become rounded as it crashed against both the valley sides and other large blocks entrained in the chaos of such a megaflood. In the absence of a glacial theory proposing a much more extensive cover of ice in the past, he put forward a mechanism to move such a large slab of rock without it losing its well defined edges. Therein lies the problem.

The sheer size of this block demands a mighty force to move it but its angularity requires a rather more passive mode of transport than would be possible in a catastrophic flood. Is such passive transit on the icebergs of a colossal flood realistic in this mountain setting? The catastrophist's megaflood does not offer a satisfactory explanation on the grounds of boulder shape. The disciples of uniformitarianism had to stretch their own imaginations to find modern analogues to explain this phenomenon. Imagine the size of the iceberg and the depth of water needed to float this block of granite! James Hutton had already provided the best explanation for the emplacement of the erratic blocks in the Alps and the adjacent lowlands, but his ideas were largely ignored or forgotten.

The diluvium debate

Large parts of northern Europe, Canada, and the northern states of the USA are covered by thick deposits of unconsolidated gravels, sands, and clays which often contain huge boulders. Many observers referred to these as superficial deposits because they were found at the land surface and typically without reaching any great depth in comparison to the hard rocks of the Earth's crust. Most geologists at this time referred to these sediments as 'the diluvium', believing they were laid down in the aftermath of a great flood. The term alluvium was reserved for modern sediments laid down by recent river activity.

The bones of now extinct fauna such as mammoths and woolly rhinos were sometimes found in the sand and gravel components of these deposits. Hans Sloane and his 18th-century contemporaries considered these to be the remains of animals that lived in the antediluvian era (the period following the Creation but before the great flood) and had perished in the deluge. For much of the 19th century there was much uncertainty about both the duration of this period and whether humans had lived alongside the animal species that were now extinct. Before we consider the deluge theory, it is worth thinking about the properties of these sediments and some of the problems they posed for the first generation of field geologists.

The deposits in Figure 9 are poorly sorted. They include sediment particles of many different sizes—in this case from very fine clays to very large boulders. This contrasts with sediments that have been laid down by rivers (alluvium) because moving water sorts sediments neatly into different size classes according to the energy of the flow. The sediments also lack obvious bedding structures with large

9. A section in boulder clay from the last glaciation in Scotland

cobbles and boulders dispersed within a fine-grained matrix of clay and silt. They are unstratified. Many of the boulders are sub-rounded indicating a degree of abrasion during transport to smooth their edges—others are more angular akin to rock particles on scree slopes.

Most observers in the early decades of the 19th century saw this heterogeneous deposit as evidence of deposition in the highly turbulent and turbid waters of a great flood. Such deposits formed just one component of a challenging geological puzzle. Today's geologists refer to such glacial deposits as boulder clay or till. In upland areas these sediments are often found in association with smoothed bedrock surfaces or down-valley of bedrock with distinctive linear furrows, scratches, or highly polished surfaces.

As with the troublesome erratic boulders, the search for a satisfactory explanation for the mode of deposition of these superficial deposits was crucial to the success of the glacial theory. But challenging the diluvialists was not easy—especially when their views were endorsed by the Church. The quest to explain these deposits produced one of the longest running debates in 19th-century geology. Were these sediments deposited by a great flood, by sediment-laden icebergs, or by glacial ice? How could such a widespread and chaotic deposit with its erratic boulders, fine sediments, and occasional fossil remains be satisfactorily explained?

William Buckland and *Reliquiae Diluvianae*

Even though Hutton's ideas gained a wide readership through the writings of John Playfair (1748–1819) in the early decades of the 19th century, many geologists opposed them vigorously because they presented a direct challenge to the Biblical version of Earth history. At this time, the Church of England held a strong influence over much of higher education and especially so in Cambridge and Oxford. The remarkable William Buckland now takes centre stage—first as champion of the Biblical Deluge and later as impassioned convert to the glacial theory and intrepid field companion of Agassiz. His is a singular contribution to an extraordinary chapter in the annals of geology.

In 1819, as the new Reader of Mineralogy and Geology at Oxford, the Reverend William Buckland delivered his inaugural lecture, 'Vindiciae Geologicae', in which he argued that the facts of the geological record were entirely compatible with the Book of Genesis. Buckland was eager to see the new discipline of geology prosper in the University so it was essential he provided this reassurance to the Oxford establishment. Buckland was an eccentric figure and a lively and engaging speaker. His lectures in Oxford were attended by prominent members of the Oxford establishment such as John Henry Newman, John Keble, and Samuel Wilberforce.

Buckland referred to the most recent and unconsolidated deposits of Britain as 'diluvial'—in other words, the product of a universal catastrophic deluge that he equated with Noah's Flood. Buckland carried out extensive fieldwork on horseback around Oxford (often wearing his academic gown), where he noticed thick piles of rounded gravels perched on hilltops. These sediments were just too high up to be laid down by modern rivers, so a catastrophic flood was invoked. He also argued for widespread 'diluvial erosion' suggesting that many of the deep valleys of England were excavated by the powerful waters of the flood. Buckland rejected Hutton's gradualist and uniformitarian ideas and saw all these features as evidence of an enormous deluge. In the final part of 'Vindiciae Geologicae' he emphatically declared:

> Again, the grand fact of a universal deluge at no very remote period is proved on grounds so decisive and incontrovertible, that, had we never heard of such an event from Scripture, or any other authority, Geology of itself must have called in the assistance of some such catastrophe, to explain the phenomena of diluvian action which are universally presented to us, and which are unintelligible without recourse to a deluge exerting its ravages at a time not more ancient than that announced in the Book of Genesis.

If the new science of geology was under suspicion in Oxford in the early part of the 19th century, this powerful affirmation of the established order of nature was surely just what the clergymen of Oxford wanted to hear. Buckland showed that geology could be a respectable companion to theology, philosophy, and the classics. Richard Chorley et al. (1964) summed up the impact of Buckland's inaugural lecture as follows:

> The 'Vindiciae Geologicae' was a sensational success and its author stood out in the popular mind as a brilliant catastrophist. He had rejected almost all of Hutton's theory and had presented a revitalized notion, that the Flood was responsible for all recent landscape features and [superficial] sediments. The diluvialists had become scientific.

Over the next few years Buckland carried out more fieldwork and strengthened his thesis on the deluge. In 1823, he published his famous *Reliquiae Diluvianae* (Relics of the Deluge). All 1,000 copies sold in a matter of months and a second edition was published the following year. Drawing upon examples from across the world, Buckland set out a detailed case for what he saw as an incontestable body of evidence for a great global inundation of very recent date. Buckland's famous work on the hyena bones that he found buried in the sediments of Kirkdale Cave in Yorkshire was a key part of his argument for a mass extinction of land animals—all wiped out by the great flood. This was 19th-century catastrophism—with geology and Genesis in harmony—in its most fully developed form.

Box 3
A geological lecture

Figure 10 shows William Buckland lecturing in the Old Ashmolean Museum in Oxford on the 15th February 1823. This historic image shows Buckland surrounded by his teaching materials—a wonderful collection of maps, drawings of geological sections, landscape paintings, and palaeontological paraphernalia spanning what we now know are millions of years of Earth history including giant ammonites from the Dorset coast and Quaternary mammals. The topic of this lecture was his discovery of the Red Lady at Paviland Cave in South Wales the previous month—a landmark in the archaeology of early humans in Britain, although we now know that it was the skeleton of a man not a woman! The skull of a mammoth was found in the same deposits. A good deal has been written about this lithograph because it was produced at a time when Buckland was trying to raise the profile of geology in the University. In this respect, attention has mainly been directed towards the people in the audience because, rather than students, they were key members of the Oxford establishment and Buckland was eager to impress them. Much less attention has been given to the fossils and maps on display from Buckland's personal collection. Several items are of particular interest here because they would become iconic symbols of the glacials and interglacials of the Quaternary ice age—although, of course, Buckland and his audience were not aware of this in 1823.

Buckland took great pride in presenting the latest information on the rapidly emerging science of geology. His lectures were highly entertaining—student satisfaction was high on his agenda. On the wall behind the rows of dons is Tilesius's drawing of our old friend the Adams Mammoth. A curved mammoth tusk is mounted on the wall above, and above that is a skull and antlers of the magnificent Irish Elk (*Megaloceros giganteus*)—another large herbivore that became extinct in Britain towards the end of the last glacial period. On the floor immediately in front of Buckland almost touching the toes of the eminent seated professor is the skull of a woolly rhinoceros. On the corner of the table facing Buckland is the characteristic curve of a hippo tusk. The hippopotamus is a distinctive feature of the Last Interglacial in Britain when the climate was warmer than today. It is not known who commissioned the lithograph of Buckland's lecture or why it was made, but it provides a unique insight into his lecturing style, his emerging collection of Quaternary fossils, and a truly formative period in the history of geology.

10. William Buckland lecturing in Oxford in 1823

Charles Lyell and uniformitarianism

Hutton's ideas on the uniformity of process and gradualism were taken up, repackaged, and widely promoted by Charles Lyell (1797–1875), who was born in the year that Hutton died. Lyell was an undergraduate at Oxford where Buckland's charismatic teaching rescued him from the tedium of a legal career. His lasting reputation was established by his three-volume *Principles of Geology* (1830–3) which ran to no less than 12 editions between 1830 and 1875. This work set out Lyell's blueprint for a new rational geology.

The principle of uniformitarianism is often expressed in the dictum 'the present is the key to the past'. This implores geologists always to call upon known physical laws and observable processes as their guide to explain the geological record. Whilst Lyell argued strongly against the diluvial theory—and spent many years gathering evidence to refute this idea—he always remained on good terms with his teacher and friend, William Buckland.

As Lyell's influence grew it became extremely unfashionable, certainly in Britain and North America, to be labelled a catastrophist. In accord with the spirit of the age, Lyell's uniformitarian creed was based on rationality and reason. He, more than anyone else in the 19th century, made geology a respectable science. The divisions between catastrophism and uniformitarianism have often been overstated, but this tension did form an intriguing backdrop to the epic debate over the glacial theory.

Lyell's drift ice theory

Lyell was the main promoter of the idea that icebergs could be the means of transport for the erratic boulders and fine sediments that made up the jumble of debris known as diluvium. He argued that during periods of warmer climate, large blocks of ice became detached from high latitudes and drifted to warmer waters carrying their loads of debris (Figure 11). As the icebergs melted,

11. Drift ice conveying large boulders and other sediments in accord with Lyell's model of marine submergence

all the sediments, large and small, were deposited subaqueously on the surface of the 'submerged continents'. Because a thick veneer of these superficial sediments covered large tracts of lowland northern Europe, marine submergence of now dry land had to be a key part of this model. This theory received strong support from many leading geologists including Roderick Murchison (1792–1871).

The Moel Tryfan shells

In June 1831, Joshua Trimmer (1795–1857) sent a letter to the Geological Society reporting the presence of beautifully preserved marine shells at 1,400 feet (c.427 m) above sea level in the superficial deposits on the top of Moel Tryfan, a small peak to the southeast of Caernarvon in North Wales about 10 km from the Irish Sea coast. The deposits also contained erratic gravels from the Lake District and southwest Scotland. Trimmer had developed a keen interest in geology whilst overseeing his father's copper mines and slate quarries in Snowdonia. His letter was read to the Society by William Buckland. Trimmer and Buckland saw these elevated shells and far-travelled erratics as further proof of a great deluge generated by a large flood wave coming from the north. After the reading of Trimmer's letter, Charles Lyell wrote to his friend, the pioneer dinosaur palaeontologist, Gideon Mantel (1790–1852):

> Murchison and his wife are gone to make a tour in Wales, where a certain Trimmer has found near Snowdon 'crag' shells at a height of 1,000 feet, which Buckland and he convey hither by the deluge.

Lyell took a rather different view. He was eager to make the case that these 'shelly drifts' reinforced the notion that the superficial sediments of the Quaternary were nothing to do with a great flood—they were laid down at a gentle pace by melting drift ice in deep water (Figure 11). These elevated shells allowed Lyell and others to invoke a great marine submergence (with the continents at a lower level) so that much of Britain and northern Europe was invaded by the sea and was therefore accessible to debris-laden drift ice.

The sturdy pontoon of ice with its neat cargo of giant boulders shown in Figure 11 probably owes a good deal to the wishful thinking of Lyell and other 19th-century geologists who worked hard to find 'a cause now in operation' to explain everything they observed in the Quaternary record. The occasional sighting of a boulder embedded in floating ice in the Atlantic generated much excitement in this era—yet much less attention was given to issues of how representative such occurrences actually were. But one thing the uniformitarians had in abundance was time, so the scarcity of boulder-laden icebergs was never a barrier to the wide acceptance of Lyell's drift ice theory.

Chapter 3
Monster glaciers

...it seemed as if Nature was stepping out of its normal course,
and the glaciers expanded rapidly...

Bernhard Friedrich Kuhn (1787)

In June 1818, a great flood swept away bridges, devastated
farmland, and destroyed 400 houses in the town of Martigny in
the Swiss Alps. Forty-seven people lost their lives in the Val
de Bagnes. The flood was caused by rapid emptying of an
ice-dammed glacial lake—the Lac de Mauvoisin—that had formed
behind a barricade of ice blocks from the avalanching Gietroz
Glacier. Fed by the flow of the River Dranse, the lake grew longer
and deeper throughout the spring of 1818. By early May the ice
barrier was almost a kilometre in length and over 120 m high. The
locals were all too aware of the developing hazard since an
outburst flood in the same valley had killed 140 people and
wrecked 500 buildings in 1595.

A young engineer, Ignace Venetz (1788–1859), was tasked to drain
the lake to eradicate the danger. Under his supervision, a trench
was cut across the ice dam to form an outlet for the rising water.
As enormous hunks of ice toppled from the snout of the hanging
glacier above, teams of men worked around the clock for much of
May and early June hacking a trench across the ice barrier. The ice
was cut with axes and hauled away on sleds. It was exceptionally

12. The Gietroz Glacier blocking the Val de Bagnes in 1818

gruelling and hazardous work. Some workers abandoned the struggle in fear for their lives, creating false alarms and chaotic evacuations in the valley below. The authorities even commissioned artists to record the scene (Figure 12).

At its greatest extent, the lake was 2 km long, about 60 m deep, with a volume of 30 million cubic metres. Despite the best efforts of Venetz and his men, the ice barrier became increasingly unstable and the trench was repeatedly choked with slabs of floating ice. On 16th June the ice dam failed, sending an immense flood wave charged with blocks of ice cascading down the Val de Bagnes. Alarms were sounded and bonfires were lit. Without the work of Venetz and his men, the number of casualties could have been much closer to the debacle of 1595.

Later that summer a young Charles Lyell (then aged 21 and reading Classics at Oxford but attending the geological lectures of William Buckland) visited the valley with his parents and two elder sisters on a family tour of the Continent. The ground floor of their hotel in Martigny had been filled with sediment from the flood. Lyell surveyed the breached ice barrier, the remnants of the lake with its blocks of floating ice, and the aftermath of the great flood. He discussed the event with eye witnesses and took detailed field notes. This flood would feature in his *Principles of Geology* when it was published 12 years later and floating blocks of ice were prominent in his thinking throughout his long and distinguished career. Lyell may have noted that large floods were observable phenomena that left distinctive signatures in the landscape. They had clearly defined boundaries and their causes could be established. This was not a universal deluge. Much later such events were easily accommodated within his uniformitarian manifesto.

The calamitous inundation of 1818 was widely reported across Europe. It is recounted here because it played a pivotal role in shifting the direction of geological thinking in the nineteenth century. Indeed, it is profoundly ironic that this catastrophic flood not only influenced Lyell's early theorizing about 'existing causes' (that would lead to the demise of catastrophism following the publication of *Principles*), but it also motivated Venetz to embark upon a careful study of glaciers and glacial landforms in the Canton of Valais that would ultimately lead, via Jean de Charpentier, Karl Schimper, and Louis Agassiz, to a new paradigm in geology—*The Ice Age*—and the end of the old diluvial theory.

The ice dam breach raised more immediate practical concerns as early 19th-century Europe was still in the grip of what we now call *The Little Ice Age*. The River Thames froze in London for the last time in 1814. Many Alpine glaciers had advanced rapidly after the unusually cold and snowy years of 1816 ('the year without a summer') and 1817. The chaotic crevassed surface of the Gietroz Glacier and its ice avalanches formed a major hazard for many years. Venetz was forced to engineer new ways to maintain the natural flow of the river.

Immediately after the disaster, Venetz was commissioned to conduct a survey of glaciers in the wider Valais region to better understand their behaviour and assess the threat they posed. He collated climate data and carried out a wide ranging programme of field observations to explore the links between climate and glacier fluctuations. With the help of Jean-Pierre Perraudin (1767–1858), a mountain guide and chamois hunter who lived in the small village of Lourtier in the Val de Bagnes, Venetz was able to take his ideas much further. Perraudin had expert knowledge of these alpine terrains and quite independently had developed his own glacial theory as early as 1815. He too had observed the traces of past glaciers in the landscape and he showed them to Venetz. Importantly, Perraudin had worked out that scratched rocks and moraine ridges located many tens of kilometres down valley from the active glaciers signalled a period when Alpine glaciers had been far larger. These distant moraines were often hidden in the Alpine forests but their significance was clear to Perraudin. His contributions were not recognized by the wider scientific community, but Venetz was generous in acknowledging Perraudin's insights. There is a small museum (*The Glacier Museum*) in Perraudin's former house in Lourtier dedicated to the glacial pioneers.

Venetz produced remarkably detailed topographic maps of lateral and terminal moraines that lay far down valley of the modern glaciers. He was able to show that many glaciers had advanced and retreated in the historical period. His was the first systematic analysis of climate–glacier–landscape interactions. For the Glacier de Rossboden near the village of Simplon in southern Switzerland he mapped the dimensions of the ancient glacier and concluded that:

> The moraines of the Rossboden glacier, on the Simplon, prove in a quite striking manner the gigantic former size of this glacier, having arrived very close to the place where the village of Simplon is today.

In 1821, Venetz presented his findings to the Société Helvétique des Sciences Naturelles, setting out Perraudin's ideas alongside his own. The paper had little impact, however, and would not see publication until 1833. But Venetz did not give up. He broadened the geographical scope of his fieldwork and gathered more evidence. In July 1829, the Société held its 15th annual meeting in the spectacular setting of the Great St Bernard Hospice at 2,469 m above sea level on the ridge between the two highest mountains of the Alps: Mont Blanc and Monte Rosa. Was there a finer location to present a grand theory of Alpine glaciation? The official report of this meeting describes Venetz's paper as follows:

> Mr Venetz read a paper on the extension that he has inferred that glaciers formerly had, and on their retreat to their present limits. He attributes the masses of blocks of Alpine rocks that are scattered at various points in the Alps and the Jura, as well as in several regions in northern Europe, to the existence of immense glaciers that have since disappeared, and of which these blocks formed the moraines. He supports this hypothesis by citing several facts that he has observed in the neighbourhood of glaciers in the Alps of the Valais.

These were remarkable insights for the late 1820s. Here was a glacial theory based on systematic observation of active glaciers and applied to the erosional and depositional signatures left by ancient ice masses—a model of Huttonian uniformitarian reasoning. Venetz had now extended the scope of his 1821 paper to challenge the diluvial and floating ice theories for erratic block emplacement across the whole of Europe. Remember that Buckland's best-selling *Reliquiae Diluvianae* (1823) was still immensely influential in Britain at this time. Unfortunately, the vast majority of the audience in the Great St Bernard Hospice were unmoved by his conclusions and they made little impression elsewhere. One member of the audience did see merit in these bold new ideas, however, and he willingly shared his own observations with Venetz.

Jean de Charpentier (1786–1855) was manager of the famous salt mines at Bex less than 20 km along the Rhône Valley from Martigny. Charpentier had studied geology under Abraham

Werner (1749–1817), the founding father of German geology, at the Freiberg Mining Academy. He paid particular attention to the disposition of large erratic blocks and the occurrence of polished and striated bedrock surfaces in the deep valleys of western Switzerland. A major step forward was Charpentier's recognition of a clear relationship between the elevation of the erratic blocks in the Rhône Valley and the vertical extent of glacially smoothed rock walls. He noted that the bedrock valley sides above the erratic blocks were not worn smooth because they must have been *above* the level of the ancient glacier surface. The rock walls below the erratics always bore the hallmarks of contact with glacial ice. We call this boundary the trimline. It is often clearly marked in hard bedrock because the texture of the valley sides above the glacier surface is fractured due to attack by frost weathering. The detachment of rock particles above the trimline adds debris to lateral moraines and the glacier surface.

These insights allowed Charpentier to reconstruct the *vertical* extent of former glaciers for the first time. Venetz and Perraudin had already shown how to demarcate the *length* and *width* of glaciers using the terminal and lateral moraines in these valleys. Charpentier described some of the most striking erratic boulders in the Alps including the colossal Pierre des Marmettes at Monthey—the largest erratic block in Switzerland that still has its own chapel on top! This hunk of grey granite came from Val Ferret on Mont Blanc over 40 km to the south.

As Charpentier mapped the giant erratics, polished bedrock surfaces, and moraines in the Rhône Valley, it became clear to him that the valley must once have been occupied by a truly enormous glacier or '*glacier-monstre*' as he called it. Charpentier met with Charles Lyell in 1832 when Lyell visited Switzerland on his honeymoon. He shared his emerging ideas of an Alpine glaciation that was more extensive than anyone had previously conceived. Like Henry de la Beche on the south side of the Alps, Lyell was highly sceptical about the role of glacial ice in transporting the large erratic boulders of the lower Rhône Valley and Jura, preferring his own model of transport upon floating blocks of ice following a dam burst flood. Even in 1832, it must have been difficult to envisage the emplacement of Charpentier's monster erratics by floating ice. It is a great pity that we don't have a record of the discussion between Charpentier and Lyell.

Charpentier and Venetz were very perceptive field scientists. They accumulated a large body of field evidence and were able to integrate details of sediment types and landforms from many localities to see the bigger picture of landscape history. In 1836, Charpentier published a key paper setting out the main findings of their glacial work in Robert Jameson's influential *Edinburgh New Philosophical Journal*:

> If I were not afraid of fatiguing the reader, I would further quote a multitude of phenomena, which, in this manner, stand also in connection with erratic blocks; and I could still add many circumstances, presented at almost every step on our mountains and in our plains, and which, taken together, support the opinion, that, in former times, all the alpine valleys, and part of the plains at the foot of the mountains, were occupied by enormous glaciers.

Figure 13 shows Charpentier's ground-breaking reconstruction with the Rhône Valley and Lake Geneva occupied by a colossal glacier that fanned out across the lowlands of the Pays de Vaude to the Jura Hills. At its maximum extent this enormous body of ice extended over 120 km from Geneva in the west to the lower reaches of the River Aare beyond Bern in the east. The active glaciers of the early 19th century are completely dwarfed by Charpentier's '*glacier-monstre*'. This elegant map was published in colour in 1841 in his *Essai sur les glaciers et sur le terrain erratique du bassin du Rhône*. For reasons that will become clear in the next chapter, this pioneering ice sheet reconstruction is commonly overlooked in accounts of the ice age debate.

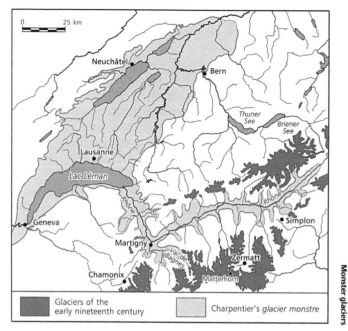

13. Charpentier's pioneering glacial reconstruction

Even before Charpentier was thinking about large ice masses in
Switzerland, Jens Esmark (1763–1839), who was born in Denmark
but became the first professor of geology in Norway, had
suggested that northern European glaciers had been much more
extensive in the past and were responsible for the transport of
large erratic boulders and the formation of moraines. Esmark also
recognized the key role of deep bedrock erosion by glacial ice in
the formation of the spectacular Norwegian fjords. He worked out
that glaciers in Norway had once extended down to sea level.

Esmark's ideas were also translated into English and published by
Robert Jameson in 1826, a decade in advance of Charpentier's
paper. Esmark discussed a large body of evidence pointing to an
extensive glaciation of northern Europe. The following extracts set

47

out some of his key conclusions and reveal how his thinking was far in advance of his contemporaries:

> In short, in every country, whether it be mountainous or flat, we shall find similar traces of the operation of masses of ice. The prominent conglomerations [moraines and other superficial Quaternary sediments] to be found in many districts, may be easily accounted for in the same manner. But it is particularly in Norway I have found many proofs of the operation of immense masses of ice which have now disappeared.

Esmark's observations about ice masses extending out into lowland countries illustrate quite how far ahead of his time he actually was. We will see in later chapters that it would be many decades before there was general acceptance within the geological community that glaciers could spread out across low gradient landscapes. His explanation for the emplacement of large erratic blocks also proved to be correct:

> In no other satisfactory way than by the operation of ice can we explain how these prodigiously large loose stones, sometimes with sharp corners, have been brought up to the ridges and tops of high mountains...

Unfortunately, even Esmark's carefully argued paper held little sway in Britain and elsewhere, although Robert Jameson (1774–1854), who was Regius Professor at the University of Edinburgh for over five decades (and tutored Darwin at Edinburgh between 1825 and 1827), did include some of Esmark's ideas in his classes. According to the lecture notes of James Forbes (1809–68), Jameson even speculated about the former presence of glaciers in Scotland—although he did not identify specific field sites and these radical ideas were never published.

Away from the mountains of Norway and Switzerland, however, in the lecture theatres and academic societies of Paris, Berlin, and

London, the geological establishment was slow to take up these ideas, even though they were published in both English and French and were widely available. Much of the debate in the 1820s and early 1830s centred on the controversy over the evolution of valleys between the fluvialists (Hutton, Playfair, and others), who advocated slow river erosion, and the diluvialists (Buckland, De la Beche, and others) who argued that big valleys and large boulders needed huge deluges. The role of glaciers in valley and fjord formation was not considered. Another key problem for Esmark, Venetz, and Charpentier was explaining quite *how* such large glaciers had formed when climate change was also considered to be out of the question.

In the 1820s and 1830s, Esmark, Venetz, and Charpentier conducted rigorous field studies and arrived at the same conclusion: vast glaciers had once existed in the mountains of Europe and extended out into the lowlands. But they did not invoke a colder climate to explain the formation of these great glaciers. Charpentier speculated that the elevation of the Alps themselves was formerly much greater so that more snowfall at higher elevations might be sufficient to promote a much more extensive cover of ice.

Venetz, Charpentier, and Esmark were the first glacial geomorphologists to attempt large-scale landscape reconstructions of former ice masses based on systematic geological fieldwork. They made important theoretical advances and they deserve their own chapter in this book. Unfortunately, because they largely worked in isolation and the most famous living geologists of the day were either satisfied with the existing diluvial explanations or engaged in other debates, their pioneering ideas were largely ignored. The key elements of a glacial theory were in place but nobody was listening. If a glacial theory was to enter the mainstream, it needed a champion.

Chapter 4
Die Eiszeit

> Imagination will often carry us to worlds that never were, but
> without it we go nowhere.
>
> Carl Sagan (1980)

Louis Agassiz

Jean Louis Rodolphe Agassiz (Figure 14) was born in Môtier in
western Switzerland in 1807, the year the Geological Society was
founded in London. After studying medicine (to please his
parents) in Zürich and Heidelberg, he returned to his boyhood
passion for natural history and was awarded his PhD in zoology in
Munich in 1829. Agassiz then based himself in Paris so that he
could work on the fossil fish at the Museum of Natural History
under the guidance of the great Baron Georges Cuvier, the most
famous naturalist in Europe at that time. Cuvier recognized
Agassiz's talent and granted him full access to the Museum's
wonderful fossil collections. Whilst in Paris, Agassiz also came
under the influence of Alexander von Humboldt, the great
Prussian geographer, naturalist, and explorer. These most
illustrious of mentors encouraged Agassiz to follow his dual
interests in zoology and geology. This he did and their influences
remained with him for the rest of his life.

14. Louis Agassiz

Shortly after Cuvier's death during the devastating cholera epidemic of 1832, Agassiz left Paris to take up a professorship at the Academy of Neuchâtel in his native Switzerland. As a protégé of Cuvier, Agassiz was *the* rising star in palaeontology and the leading authority on the anatomical and taxonomic investigation of fossil fish. He travelled widely in Europe carrying out fieldwork, delivering lectures, and inspecting fossil collections. Agassiz visited Mary Anning (1799–1847) in Lyme Regis in 1834 to study the treasures in her collections and to hunt for fresh specimens in the Jurassic cliffs. He corresponded with the leading natural scientists of the day, including, in Britain, Buckland, Lyell, and Sedgwick and, across the Atlantic, with Benjamin Silliman in Yale who founded the *American Journal of Science* in 1818. Agassiz's expertise was in great demand.

In 1835, Agassiz turned his attention to the long-standing controversy surrounding the origin of the great granite erratics on the Jura. One of the most famous, the Pierre à Bot, sat on the hillside just outside Neuchâtel. Agassiz was acquainted with the work of Venetz and Charpentier but, like the great majority of his contemporaries, he was not convinced by their glacial theory. However, Agassiz was eager to assess the evidence himself. So in the summer of 1836, as a guest of his friend Charpentier, Agassiz spent several weeks touring the Rhône Valley scrutinizing the evidence for these ancient enlarged glaciers.

Charpentier and Venetz taught him how to read the signs of ancient ice: the polished bedrock, the hanging valleys, the distant, degraded moraines. They enthusiastically shared their ideas about an ice-bound past. Agassiz was sceptical at first, but soon developed a fanatical interest in glaciers and glaciated terrains. He began to see the landscape in a new light. Agassiz then initiated his own phase of glacier observations and started to think deeply about ice and life in the past.

The Neuchâtel Discourse

The following summer the Swiss Academy of Natural Sciences held its annual symposium in Neuchâtel. Its members were expecting to hear about their young President's latest work on ancient fish. Instead, Agassiz stunned them by setting out a grand theory of glaciation, ice age epochs, and the history of life. To prepare for his lecture Agassiz had borrowed some notes from his colleague Karl Friedrich Schimper (1803–67), a botanist and mutual friend of Charpentier who had long-standing interests in glacial phenomena and novel ideas about long-term changes in climate. Agassiz was an accomplished and charismatic speaker. He began by dismissing existing explanations for the transport of erratic blocks to the Jura: discounting diluvial currents, floating ice, and catastrophic mudflows. He drew heavily on the work of

Esmark, Venetz, and Charpentier and added his own grand embellishments:

> The appearance of the Alps, the result of the greatest convulsion which has modified the surface of our globe, found its surface covered with ice, at least from the North Pole to the shores of the Mediterranean and Caspian Seas.

Agassiz had taken Charpentier's *glacier-monstre* and spread it over much of Europe, northern Asia, and North America. It was a startling proclamation that left his audience flabbergasted. This great ice sheet, he argued, once enveloped much of the northern hemisphere—the modern glaciers of the Alps were seen as the shrivelled remnants of this vast shield of ice. Interestingly, Agassiz argued that the erratic blocks on the Jura had slid across this sloping ice surface to the flanks of the Alps and had not been transported there by moving glacier ice.

In another key departure from Charpentier's thesis, Agassiz was certain that this glacial world was associated with an interval of much colder climate. Agassiz firmly rejected the prevailing idea that Earth history involved only gradual long-term cooling:

> The admission of an epoch of cold, which was so intense as to cover the Earth to such distances from the poles with so vast a mass of ice as we have been contemplating, is a supposition which appears in direct contradiction with those well known facts, which show a considerable cooling of the Earth since the most remote period.

Stimulated by his discussions with Schimper, he argued instead that this general cooling trend was punctuated by a *series* of ice epochs of intense cold. Agassiz also saw these epochs of intense freezing as a means of resetting the biological clock. He merged his geological and zoological musings to come up with a new theory to explain the major changes in life forms revealed in the fossil record. He suggested that ice epochs were responsible for wiping out 'previous creations'. The present creation, he argued, was preceded by a great ice epoch, not a great flood. Cuvier, the arch-catastrophist, would surely have been impressed. Agassiz declared:

> The epoch of intense cold which preceded the present creation has been only a temporary oscillation of the Earth's temperature, more important than the century-long phases of cooling undergone by the Alpine valleys. It was associated with the disappearance of the animals of the diluvial epoch of the geologists, as still demonstrated by the Siberian mammoths; it preceded the uplifting of the Alps and the appearance of the present-day living organisms, as demonstrated by the moraines and the existence of fishes in our lakes.

One of the most distinguished geologists of the day, Leopold von Buch (1774–1853), was in the audience in Neuchâtel. Von Buch had spent many years working on the rocks of the Jura (which gave its name to the Jurassic Period that he defined) and saw it as *his* domain. Like Lyell, he had been inspired by the Val de Bagnes flood of 1818. Von Buch had argued that the Jura erratics were transported by a violent mudflow from the steep basins of the Alps. Agassiz had just dismissed this theory along with all the others. Von Buch did not approve of the glacial theory in any form—he did not enjoy Agassiz's paper.

Agassiz delivered his famous lecture on 24th July 1837. Despite its hostile reception, he pursued his new interest with the zeal of a convert. In stark contrast to his colleague Schimper, who published very little, Agassiz wasted no time in getting his work into print. It has been said that Agassiz wrote his *Discours de Neuchâtel* in a single evening. The written version was published later that year in the 22nd volume of the *Actes de la Société Helvétique des Sciences Naturelles* (1837). Schimper published a short note at the end of the same volume. Its title marks his most enduring contribution to geology. It was Schimper who coined the term *Die Eiszeit* (The Ice Age): it was here that it appeared in print for the first time.

News of Agassiz's great ice age spread across Europe creating quite a stir. At the end of 1837, von Humboldt wrote a long letter cautioning his young friend against working too hard and spreading himself too thinly. The letter was cordial and supportive, but it set out the view then held by many naturalists who could not accept the notion of a former epoch of cold climate. Like his old friend von Buch, von Humboldt was very uncomfortable with the notion of a global ice age when, he contended, the evidence pointed to a warmer, *not colder*, climate in the past:

> With reference to the general or periodical lowering of the temperature of the globe, I have never thought it necessary, on account of the elephant of the Lena [the Adams Mammoth], to admit that sudden frost of which Cuvier used to speak…
>
> It is a slight local phenomenon. To me, the ensemble of geological phenomena seems to prove, not the prevalence of this glacial surface on which you would carry along your boulders, but a very high temperature spreading almost to the poles, a temperature favorable to organizations resembling those now living in the tropics. Your ice frightens me…

A leading climate change sceptic of the period, von Humboldt was eager to quash this notion of an ice age climate. He signed off by urging Agassiz to get back to what he knew best and to focus on his fossil fish:

> No more ice, not much on echinoderms, plenty of fish…

Agassiz was not to be dissuaded. With his burning ambition and prodigious appetite for work, he settled into a seasonal routine of glaciers in the summer and fossil fish in the winter. The summers of 1837, 1838, and 1839 were spent observing the glaciers of the Swiss Alps, gathering data for a major monograph. In 1838,

Agassiz's Neuchâtel paper was published in English in Jameson's *Edinburgh New Philosophical Journal*, 12 years *after* the publication of Esmark's landmark paper in the same journal.

Whilst Venetz, Charpentier, and Schimper were accorded some acknowledgement in his early publications, the spotlight was now firmly fixed on Agassiz as the primary advocate of the glacial theory. He was determined to ensure that *his* glacial epoch would not become just another footnote in the history of geology. Over the next few years, as Agassiz's celebrity grew, his old friends became increasingly embittered as they were frozen out of the picture. Much later, in the 1880s, there were efforts to secure greater recognition for Schimper's contribution to the idea of an ice age past.

The first British geologist to respond to Agassiz's conversion was his friend and fellow fossil hunter William Buckland. In the summer of 1838, Buckland travelled to Switzerland with his wife to inspect the evidence with his own eyes. Mary Buckland wrote to Agassiz from Interlaken:

> We have made a good tour of the Oberland and have seen glaciers, etc., but Dr. Buckland is as far as ever from agreeing with you.

Despite his initial reservations, Buckland became utterly convinced by the field evidence for a much greater extent of ice in the past. Though it challenged many of his long held views on diluvial phenomena, Buckland embraced this new paradigm and immediately began to think of features in his native land that might be better explained by glacial processes. Agassiz showed him the grooved and polished rocks, the erratic blocks perched high on the valley sides, and the distinctive terminal moraines many kilometres down valley from the nearest glacier. Agassiz converted Buckland just as Charpentier and Venetz had converted him. Buckland became an ardent supporter of the glacial theory. At the earliest opportunity, he and Agassiz agreed to seek out traces of a glacial past in the uplands of Britain.

Études sur les glaciers

Agassiz published his famous *Études sur les glaciers* at the end of 1840. This was the first major scientific publication that argued for the existence of an ice age in Earth history. Esmark, Venetz, and Charpentier had made the case for a much greater expanse of ice cover in the past, but they did not associate this with a glacial epoch of intense cold climate. Agassiz's monograph was lavishly illustrated with a wonderful series of lithographic prints. Figure 15, for example, shows the upper section of the debris-laden Zermatt Glacier next to a broad bench of ice-moulded bedrock. As the first atlas of glaciers and glacial landforms, it became a remarkably influential work. There were chapters on glacier formation, glacier movement, and bedrock erosion and the transport of boulders.

In the final chapter, Agassiz expanded on his *Discours de Neuchâtel* setting out his latest thinking on large-scale glaciation and faunal extinction:

> The Earth was covered by a huge ice sheet which buried the Siberian mammoths, and reached just as far south as did the phenomenon of erratic boulders. This ice sheet filled all the irregularities of the surface of Europe before the uplift of the Alps, the Baltic Sea, all the lakes of Northern Germany and Switzerland. It extended beyond the shorelines of the Mediterranean and of the Atlantic Ocean, and even covered completely North America and Asiatic Russia.

Agassiz was occasionally prone to exaggeration and he did get rather carried away with the proposed extent of his ice sheet. This irritated Charpentier intensely because, in marked contrast, the limits of Charpentier's *glacier-monstre* were carefully constrained and based on many years of geological observations. Agassiz had never visited Siberia or North America and there was certainly no geological evidence to suggest that an ice sheet had reached

15. The Zermatt Glacier from *Études sur les glaciers*

anywhere close to the Mediterranean Sea. But Agassiz was brilliant at self-promotion and never timid in putting forward bold ideas. He had turned the glacial theory into a global catastrophe. Adam Sedgwick offered the following view on Agassiz's monograph:

> I have read his ice book. It is excellent, but in the last chapter he loses his balance, and runs away with the bit in his mouth.

Agassiz's book preceded Charpentier's own glacial monograph by a matter of months, the latter finally reaching publication in 1841. Quite understandably, Charpentier, who with Venetz was the source of most of the original ideas about the former extension of Alpine glaciers, was less than pleased that Agassiz had beaten him into print. It was he, after all, who had invested years of work into his glacial research long before Agassiz had shown the slightest interest. From 1840, Agassiz became the main promoter of the glacial theory and it was his name that became indelibly linked to it.

Key strands of Agassiz's grand ice age theory had their origins in two pivotal events: the study of the Adams Mammoth and the Val de Bagnes flood. The first led to Cuvier's critical conclusions on the Siberian mammoths. The second led directly to the first large-scale reconstruction of an Alpine ice sheet. Agassiz was intimately associated with the architects of both. He was therefore perfectly placed to meld these ideas with his own and with the musings of Schimper to create a new grand theory. For the first time a glacial theory had momentum and a famous name with impeccable credentials attached to it. Agassiz was charismatic, hugely energetic, and well connected. He also had the means to travel to international meetings. Even if people didn't like his audacious ice age theory, they would now have to sit up and take notice.

Chapter 5
1840

Ideas without precedent are generally looked upon with
disfavour and men are shocked if their conceptions of an
orderly world are challenged.

J Harlen Bretz (1928)

The year 1840 was the year the ice age arrived in Britain. As
President of the Geological Society (1839–41), Buckland invited
Agassiz to present his glacier work in London. Agassiz travelled to
Britain in September to speak first at the British Association
Meeting in Glasgow. It was there that Agassiz first saw evidence
for the former presence of glaciers in Britain as he inspected the
Quaternary deposits exposed on a building site during a walk
around the city. Glasgow sits on a field of drumlins. Immediately
after the Glasgow meeting, Agassiz undertook his famous tour of
Scotland and Ireland looking for evidence of glaciation. Buckland
was his guide for much of the Scottish leg. Their itinerary has
been reconstructed by Gordon Davies in the *Annals of Science*
(1968). Never one to waste an opportunity, Agassiz appears to
have spent as much time looking at fossil fish collections and
exposures in the Old Red Sandstone as he did searching for the
signs of ancient ice.

Agassiz and Buckland left Glasgow on 23rd September travelling
north by stagecoach to the Highlands. They passed Loch Lomond

and Loch Fyne en route to Inveraray, where Agassiz famously declared to Buckland:

Here we shall see our first traces of glaciers.

Many years later Agassiz recalled how their stagecoach actually drove over a terminal moraine as they entered the valley. This must have been a tremendously exhilarating time for Buckland and Agassiz. Both were animated and compulsive conversationalists who knew all the big fish in their 19th-century world of natural science. They were fully aware of the significance of their journey of discovery. Geology had a new theory and Scotland was the birthplace of Hutton, Playfair, Lyell, and Murchison. They were about to show the world that the Great Glens of Scotland had been carved by mighty glaciers.

They recorded glacial features all the way to Fort William. At every turn they saw polished and scratched rocks, erratic boulders, U-shaped valleys, glacial lakes, lateral and terminal moraines. Agassiz and Buckland were in no doubt that this landscape had been sculpted by ice. When they made their way to the glens around Ben Nevis and to the famous Parallel Roads of Glen Roy, Agassiz realized immediately that Darwin's paper of the previous year was wrong. The Parallel Roads were former shorelines of a lake that had been dammed by a large glacier filling Glen Spean. The blockage of the Val de Bagnes by the Gietroz Glacier in 1818 provided the perfect analogue. Agassiz produced a map showing the position of the moraines and other glacial features (Figure 16).

Never slow to promote his work and acutely aware of the importance of getting these revelations into print at the earliest opportunity, Agassiz wrote to Robert Jameson on 3rd October from Fort Augustus on the edge of Loch Ness. The letter came too late to be included in the latest issue of Jameson's *Edinburgh New Philosophical Journal*, so Jameson sent it to Charles Maclaren, a journalist and geologist who co-founded *The Scotsman* newspaper

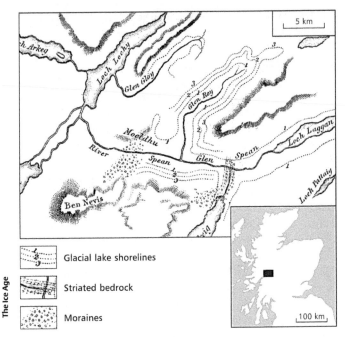

Glacial lake shorelines	
Striated bedrock	
Moraines	

16. A map of Glen Roy from Agassiz's 1842 paper

in 1817. Maclaren fully appreciated the significance of these findings and the potential for a journalistic scoop.

The extract from Agassiz's letter, as it appeared in *The Scotsman*, is shown in Figure 17. It was certainly not conventional to report geological research in the newspapers, but Agassiz was breaking new ground in more ways than one. In the first-person narrative of this letter we find Agassiz speaking about himself reporting *his* discoveries. It is a pity that he did not see fit to mention Buckland in this correspondence. Putting issues of priority to one side for now, this may well have been the first article on climate change to appear in a newspaper; it was the first to announce The Ice Age to the general public. Over the next few years Charles Maclaren

DISCOVERY OF THE FORMER EXISTENCE OF GLACIERS IN SCOTLAND, ESPECIALLY IN THE HIGHLANDS, BY PROFESSOR AGASSIZ.

Extract from letter from *Professor Agassiz* to Professor Jameson, dated Fort Augustus, Oct. 3d ; received Oct. 6th :—" After having obtained in Switzerland the most conclusive proofs, that at a former period the glaciers were of much greater extent than at present, nay, that they had covered the whole country, and had transported the erratic blocks to the places where these are now found, it was my wish to examine a country where glaciers are no longer met with, but in which they might formerly have existed. I therefore directed my attention to Scotland, and scarcely arrived in Glasgow, when I found remote traces of the action of glaciers, and the nearer I approached the high mountain chains these became more distinct, until, at the foot of Ben Nevis, and in the principal valleys, I discovered the most distinct *morains* and polished rocky surfaces, just as in the valleys of the Swiss Alps, in the region of existing glaciers; so that the existence of glaciers in Scotland at early periods can no longer be doubted. The parallel roads of Glen Roy are intimately connected with this former occurrence of glaciers, and have been caused by a glacier from Ben Nevis. The phenomenon must have been precisely analogous to the glacier-lakes of the Tyrol, and to the event that took place in the valley of Bagne. It appeared to me that you would be glad to be able to announce, for the first time, in your extensively-read journal, the intelligence of the discovery of so important a geological fact."

17. Agassiz's letter in *The Scotsman* in October 1840

published a regular series of articles in *The Scotsman* on glacial processes and landforms.

Why did Agassiz write this letter? Why did he fail to mention the important contributions of his companion William Buckland? Agassiz also takes *all* the credit for the work in the Alps—the pioneering efforts of Venetz and Charpentier have been air-brushed away. It is important to point out that Buckland had already recognized evidence for glacial activity in Scotland *before* Agassiz arrived in the country. Buckland's many glacial discoveries have been documented by Patrick Boylan, Emeritus Professor at City University, London. If priority was not so important to Buckland, it certainly was to Agassiz. It was Agassiz who came up with the best explanation for the Parallel Roads of Glen Roy, but Buckland must take the credit for being the first to recognize evidence for glaciation in Scotland, England, and Wales.

Agassiz continued his search in Ireland whilst Buckland headed to Kirriemuir near Forfar to pay a visit to Charles Lyell at his grand country estate of Kinnordy House. It was there that Buckland showed his old pupil the moraines on his estate and convinced

1840

Lyell that glaciers had once existed in Scotland. On 15th October Buckland wrote to Agassiz from the splendour of the Marquess of Breadalbane's Taymouth Castle in the Scottish Highlands (Buckland was extremely well connected). He could barely hide his glee:

> ...Lyell has adopted your theory *in toto*!!! On my showing him a beautiful cluster of moraines, within two miles of his father's house, he instantly accepted it, as solving a host of difficulties that have all his life embarrassed him. And not these only, but similar moraines and detritus of moraines, that cover half of the adjoining counties are explicable on your theory, and he has consented to my proposal that he should immediately lay them all down on a map of the county and describe them in a paper to be read the day after yours at the Geological Society.

This was an immensely significant coup for Buckland. Lyell was now a hugely influential figure—if the glacial theory was to cut any ice with the members of the Geological Society in London, it was crucial to get him on board. This extract also shows just how successfully Agassiz had commandeered the glacial theory and made it his own.

Agassiz travelled on to Ireland to visit Lord Cole (1807–86), the Third Earl of Enniskillen, who had one of the finest collections of fossil fish in the British Isles. He noted moraines and polished bedrock in various parts of Ireland before returning to Scotland to meet Robert Jameson and Edward Forbes in Edinburgh.

Following the article in *The Scotsman*, Agassiz had become quite a celebrity. Like some travelling geological guru, he was escorted to various bedrock outcrops in the city and asked to inspect them for traces of glacial activity. The story goes that when he saw striations on the rocks on the southern side of Blackford Hill, Agassiz tossed his hat in the air and proclaimed: 'That is the work of ice!' This location became known as Agassiz Rock and a plaque was placed there in 1993 by Scottish National Heritage to commemorate this event.

Agassiz, Buckland, and Lyell at the Geological Society

The next chapter in a remarkable year in the history of British geology was to be played out in Somerset House in London at the Geological Society. The year of 1840 was noteworthy for other reasons too. In February, Queen Victoria had married her German cousin Prince Albert of Saxe-Coburg and Gotha. In early summer, the granite foundations for Nelson's Column were laid in Trafalgar Square and the largest brick structure in the world, Stockport Viaduct, traversing the Mersey Valley, was completed. Were these portents of an opportunity to forget old rivalries, build bridges, and embrace new ideas from Europe? How would the glacial trinity of Agassiz, Buckland, and Lyell fare in the lion's den of the Geological Society?

Agassiz read his paper on 'Glaciers and the Evidence of their having Once Existed in Scotland, Ireland, and England' on 4th November. It was followed by Buckland's 'Memoir on the Evidences of Glaciers in Scotland and the North of England'. Buckland continued this paper on 18th November when Charles Lyell presented 'On the Geological Evidence of the Former Existence of Glaciers in Forfarshire'. All three participated in the lively discussions that followed these papers.

It was not normal practice at the Geological Society to make a record of the discussion that followed a paper. However, when Agassiz, Buckland, and Lyell delivered their papers at the close of 1840, a young sub-curator at the Society, Samuel Woodward (1821–65), took it upon himself to take detailed notes of the exchanges. His record provides a fascinating insight into the immediate reception of the glacial theory by the leading British geologists. Woodward's notes were only published posthumously, over four decades later, when all the key protagonists were no longer alive. They were also published in full in 1907 when his son, Horace Bolingbroke Woodward (1848–1914), published *The History of the Geological Society of London* to mark its centenary.

Samuel Woodward's notes show that all three papers met with a rather chilly reception. Most members kept their counsel, perhaps out of respect for the eminent trio of speakers. The remarkable polymath William Whewell (1794–1866), soon to be Master of Trinity College, Cambridge (and inventor of many terms including *scientist, catastrophism,* and *uniformitarianism*), raised objections; as did George Bellas Greenough (1778–1855), one of the founding members of the Society. The record shows that Roderick Murchison provided the most vociferous and lengthy objections to the glacial theory. When considering the evidence for glacial scratches on bedrock surfaces, Murchison alluded to the scratches produced by horse-drawn carriages on the cobbled streets of London, when he famously complained:

> Are glaciers the cause? Could they be done by ice alone? If we apply it to any as the necessary cause, the day will come when we shall apply it to all. Highgate Hill will be regarded as the seat of a glacier, and Hyde Park and Belgrave Square will be the scene of its influence.

Sir Roderick Impey Murchison, 1st Baronet (1792–1871), was one of the most important geologists of the 19th century. He was a towering figure at Somerset House; it would be difficult to overstate his influence at the Geological Society and in wider scientific circles. Murchison's ground-breaking monograph *The Silurian System* (1839) had been published only the year before. It included a chapter on the superficial (Quaternary) deposits in the borderlands between England and Wales—Murchison argued forcefully for the agency of drifting ice in their deposition. The suggestion of an ice sheet covering Britain was never going to be popular with Murchison and he emerged as perhaps the most vigorous British opponent of the glacial theory.

What would the Geological Society make of their President's conversion to a new glacial theory? Buckland was a notorious eccentric, even by Victorian standards. He could get away with this in Oxford, where he was much loved by his students, but his manner was not to everyone's taste—especially in the rarefied atmosphere of gentlemen's societies in London. In his autobiography, Darwin had this to say about Buckland:

> All the leading geologists were more or less known by me, at the time when geology was advancing with triumphant steps. I liked most of them, with the exception of Buckland, who though very good-humoured and good-natured seemed to me a vulgar and almost coarse man. He was incited more by a craving for notoriety, which sometimes made him act like a buffoon, than by a love of science.

Whether or not others held this view, it is not difficult to see why the geological establishment was reluctant to back Buckland's latest project. As the most famous and vociferous of the diluvialists, Buckland had jumped ship, exchanging his universal deluge for great glaciers. His ally in this enterprise was a French-speaking Swiss zoologist—a leading authority on fossil fish who had a passing resemblance to Napoleon. Agassiz had the audacity to pitch up at the Geological Society at the end of 1840 to tell the leading British geologists of the day how they had all fundamentally misinterpreted the geology of their own backyard.

In addition, members of the Society were all too aware of Darwin's recent paper on the Parallel Roads and how this would be fatally undermined if they adopted Agassiz's theory. This much was clear in Agassiz's letter in *The Scotsman*. All this put Lyell in a rather awkward position. Darwin was a close friend and a popular and hard-working Secretary of the Society; his only substantial piece of geological fieldwork in Britain would become the first casualty of this new theory. Lyell's flirtation with the land-ice theory proved to be rather short-lived.

Murchison's backing of the drift ice theory is especially significant and many members would have thought it odd that Lyell, as the chief architect of that theory, was now advocating a role for land ice in Scotland. Murchison, the former army officer, was a tall man with a commanding presence; he was absolutely not a man to yield an inch in any argument. He was an enforcer with a reputation for 'extreme disputatiousness' who battled hard to 'vanquish those who threatened his realm'. It was a brave man who took him on at the Geological Society. So when he took the Chair in 1840 as the discussion of the papers by Agassiz, Buckland, and Lyell began, the smart money was *not* on the glacial theory.

William Buckland in North Wales

Buckland continued his fieldwork in North Wales in 1841 seeking out the evidence for glaciation in Snowdonia. He was accompanied by his friend Thomas Sopwith (1803–79), a mining engineer and railway surveyor from Newcastle-upon-Tyne, and an ardent early British convert to the glacial theory. Sopwith was also a talented cabinet maker (he was to win a prize at the Great Exhibition of 1851), renowned for his exquisite three-dimensional working models of geological phenomena. His grandson's aviation company manufactured the Sopwith Camel during the First World War.

As the highest mountain in England and Wales, the slopes and valleys around Snowdon (1,085 m) were obvious candidates to search for evidence of past glacial activity. Buckland was not disappointed. He was the first to recognize the signs of ancient glaciers in Wales. The evidence was especially impressive in the dramatic Nant Francon pass. This deep U-shaped glacial trough is fringed by cirques and hanging valleys. The old hard rocks of North Wales that Adam Sedgwick had studied for so long were ideal for registering the grooves and scratches from the passage of glacier ice. Darwin's Cwm Idwal is one of the classic cirque basins that feed the upper end of the Nant Francon valley.

Sopwith and Buckland stayed in the Goat Hotel in Bedgellert where Buckland wrote in the guest book:

> Notice to geologists—At Pont-aber-glass-llyn, 100 yards below the bridge, on the right bank of the river, and 6 m above the road, see a good example of the furrows, flutings, and striae on rounded and polished surfaces of the rock, which Agassiz refers to the action of glaciers. See many similar effects on the left, or southwest, side of the pass of Llanberis.

Remarkably, this note, signed by Buckland, hung on the wall of the Goat Hotel as recently as the 1970s, but is sadly now lost. Buckland's recognition of glacial features in North Wales inspired Darwin to revisit the landscape a decade after his initiation to geological fieldwork with Adam Sedgwick. He became fully convinced by the signs of glacial action in the mountains of Snowdonia, but he also saw evidence for a marine submergence and floating ice. Darwin published his new observations in 1842:

> We have...the clearest proofs of the existence of glaciers in this country; and it appears that when the land stood at a lower level, some of the glaciers, as in Nant-Francon, reached the sea, where icebergs charged with fragments would occasionally be formed. By this means we may suppose that the great angular blocks of Welch rocks, scattered over the central counties of England, were transported.

Buckland was delighted to hear about Darwin's paper, even if it only went so far as to countenance glacial action in the uplands. Darwin's conversion was another major coup for the glacialists and Buckland wasted no time writing to Agassiz in July 1842 after the meeting of the British Association in Manchester:

> You will, I am sure, rejoice with me at the adhesion of C. Darwin to the doctrine of ancient glaciers in North Wales, of which I send you a copy, and which was communicated to me by Dr. Fitton, during

the late meeting at Manchester, in time to be quoted by me versus Murchison, when he was proclaiming the exclusive agency of floating icebergs in drifting erratic blocks and making scratched and polished surfaces. It has raised the glacial theory fifty per cent, as far as relates to glaciers descending inclined valleys; but Hopkins and the Cantabrigians [Cambridge scholars] are still as obstinate as ever against allowing the power of expansion to move ice along great distances on horizontal surfaces…

William Hopkins FRS (1793–1866) was another fervent opponent of the glacial theory when it was first proposed in Britain. He was a farmer turned Cambridge mathematician and first class cricketer who became interested in geology in the 1830s under the influence of Adam Sedgwick. Hopkins pioneered the application of mathematics and mechanics to explore thorny geological problems such as the nature of glacier motion and the structure of the Earth's interior. He tutored William Thomson (Lord Kelvin) in Cambridge in 1845. Hopkins' initial objection to the glacial theory was based on his belief that glacier ice could not move over gently sloping terrain beyond the confines of a valley. Although he modified this position in later years, it helped to sustain the floating ice model for the deposition of drift in the lowlands. Hopkins succeeded Lyell as President of the Geological Society in 1851. He died in a lunatic asylum in Stoke Newington in 1866.

Agassiz's huge ice sheet was a non-starter for British geologists and Buckland knew it. There was fierce opposition to both the idea of land ice extending out across the lowlands and a recent glacial epoch with an arctic climate. For many, Lyell's model of sediment deposition by floating icebergs during a period of marine submergence was a perfectly rational explanation for the drift deposits of lowland Britain.

After the initial euphoria surrounding Agassiz's visit in 1840 and the hostile reaction of the Geological Society, a workable compromise emerged that satisfied most British geologists. As more systematic studies of the uplands were carried out, it became increasingly difficult to deny the evidence for the action of glaciers in Wales, Scotland, and the Lake District; many former problems became more easily reconciled with a glacial origin. The lowlands, however, were a very different matter and a large ice sheet was still out of the question. Murchison often used 'glacialist' as a derogatory label in the same way that 'catastrophist' might have been applied to someone who had suggested a wild and fantastic theory. The notion of a great marine submergence or 'glacial sea' covering much of Britain and Europe persisted into the next century. The ice age had reached Britain, but, for the time being, only in the mountains.

The death of Dean Buckland

In 1845, William Buckland was appointed Dean of Westminster by Prime Minister Sir Robert Peel (1788–1850). As head of the chapter of Westminster Abbey, the demands of this role left him with less time for geological matters. The main British advocate of the glacial theory now had to devote most of his energies elsewhere. In late 1849, Buckland contracted a debilitating neck and brain condition that prevented him from working. It plagued him for the rest of his life. The Reverend William Buckland, inspirational teacher, pioneering palaeontologist, indefatigable field geologist, the first British convert to the glacial theory, and the first to recognize evidence for glaciation in Britain, died on 14 August 1856, aged 73. Later that month he was buried in the churchyard of St Nicholas at Islip, just north of Oxford, under a block of Scottish granite.

Chapter 6
Ice sheets or icebergs?

> Concerning the Glacial period, geologists hold the most varied
> opinions, both with regard to its origin and to the mode of
> action of the ice.
>
> Thomas Belt (1877)

In February 1841, the Manx-born palaeontologist and naturalist
Edward Forbes (1815–54) wrote to Louis Agassiz and confidently
announced:

> You have made all the geologists glacier-mad here and they are
> turning Great Britain into an ice-house.

This assessment was penned just two months after the
celebrated sessions at the Geological Society. It proved to be
rather optimistic. There is no doubt that Agassiz had aroused
much British interest in glaciation—the brilliant Scottish
physicist James Forbes, for example, was inspired to begin his
own ground-breaking research on Alpine glaciers. But the
opposition to Agassiz's ice age was substantial and it came from
some of the foremost geologists of the day. It would be decades
before a majority accepted that vast tracts of Eurasia and North
America had once been covered by mighty ice sheets. Why did
the glacial theory stall after 1840?

One reason, of course, was because the glacial dream team of Agassiz, Buckland, and Lyell effectively disbanded and went their separate ways. After the bruising sessions at Somerset House, Lyell lost much of his appetite for glaciers and went back to promoting icebergs as the dominant mechanism of drift transport and deposition. Buckland's appointment as Dean of Westminster in 1845 diverted a good deal of his energies away from geology and, in 1846, after a highly successful lecture tour, Agassiz moved permanently to the United States and accepted a chair at Harvard in 1848. Whilst Agassiz continued to promote his ice age in North America with great vigour, the most high profile advocates of the glacial theory could no longer present a united front. Indeed, as Lyell's influence grew in North America, the ice sheets versus icebergs debate was played out on both sides of the Atlantic. Agassiz and Lyell found themselves on opposite sides of a new dispute that would not be settled until the end of the century.

Ice sheets or icebergs?

Most geologists in 1840 saw Agassiz's great ice sheet as a retrograde step. It was just too catastrophist—a blatant violation of hard-won uniformitarian principles. It was the antithesis of the new rational geology and was not underpinned by carefully assembled field data. So, for many, as an explanation for the superficial deposits of the Quaternary, it was no more convincing than the deluge.

The middle decades of the 19th century also had their fair share of climate change deniers. Ancient climates were supposed to be warmer not colder. The suggestion of a freezing glacial epoch in the recent geological past, followed by the temperate climate of the present, still jarred with the conventional wisdom that Earth history, from its juvenile molten state to the present, was an uninterrupted record of long-term cooling without abrupt change. Lyell's drift ice theory also provided an attractive alternative to

Agassiz's ice age because it did not demand a period of cold glacial climate in areas that now enjoy temperate conditions. Lyell had made this clear in his Presidential Address to the Geological Society in February 1836:

> It is therefore by no means necessary to speculate on the former existence of a climate more severe than that now prevailing in the Western Hemisphere in order to explain how the travelled masses in Northern Europe may have been borne along by ice.

In 1840, Lyell was struggling to reconcile the new glacial theory of Agassiz and Buckland with his own thoughts on drift ice. He was a reluctant participant in Buckland's new project and there was still strong backing for his own iceberg theory. If anything, the 1840 sessions at the Geological Society had galvanized support for floating ice as a mechanism for drift deposition in the lowlands. Lyell's model proved to be remarkably resilient—its popularity proved to be the major obstacle to the wider adoption of the land ice theory. In the 9th edition of *Principles* (1854) Lyell was still convinced that floating ice could produce many of the erosional and depositional features attributed to land-based glaciers:

> There can be little doubt that icebergs must often break off the peaks and projecting points of submarine mountains, and must grate upon and polish their surface, furrowing or scratching them in precisely the same way as we have seen that glaciers act on the solid rocks over which they are propelled.
>
> To conclude: it appears that large stones, mud, and gravel are carried down by the ice of rivers, estuaries, and glaciers, into the sea, where the tides and currents of the ocean, aided by the wind, cause them to drift for hundreds of miles from the place of their origin.

As long as Murchison and others backed this model it would not easily be displaced, especially when many refused to believe that glacier ice could advance across gently sloping lowland terrain. This was a reasonable objection at this time since the ice sheets of

Greenland and Antarctica had not yet been investigated from a glaciological point of view. It is not difficult to understand why many British geologists rejected the glacial theory when the proximity and potency of the sea was so obvious and nobody knew how large ice sheets behaved.

It is important to note that only icebergs that have calved from land-based glaciers will contain significant volumes of sediment. Frozen sea water (sea ice) does not contain sediment and cannot play a role in the deposition of drift. So did the work of *land* ice (glaciers and ice sheets) provide the best explanation for the origin of phenomena such as grooved and scratched bedrock surfaces, spreads of erratic boulders, and thick accumulations of boulder-laden drift? Did the work of continental ice sheets harmonize all the facts? Or could these features be better explained by the work of floating icebergs over submerged continents? The glacial debate had polarized around these two mechanisms (Figure 18). It was around this time that the glacial debate began in earnest on the other side of the Atlantic. We know now that North America once could boast the largest expanse of glacial terrain on Earth. How did its geologists receive the new ideas from the Alps? Would they follow Agassiz or Lyell?

18. Land ice or drift ice?

The ice age in North America

News of Agassiz's glacier studies and grand ice age theory quickly spread to North America where they generated much excitement and took some authors by surprise. Clergyman and pioneering geologist Edward Hitchcock (1793–1864) published his popular text book *Elementary Geology* in 1840, then speedily produced a second edition in 1841 when he realized the significance of the new ideas from Europe:

> The call for a second edition of this work so early has been unexpected. But I have exerted myself to adapt it to the advancing state of the Science. The most important addition related to the subject of Glaciers and Glacial Action, which is now exciting so great an interest in Europe . . . I have been favoured with an early access to the recent work of Agassiz, entitled *Études sur les glaciers*, and to some papers recently read before the London Geological Society by Agassiz, Buckland, and Lyell, on the same subject.

Hitchcock became Professor of Natural Theology and Geology at Amherst College and was one of the first Americans to publicly embrace Agassiz's ideas. He welcomed the glacial theory in the inaugural presidential address of the Association of American Geologists in Philadelphia in April 1840, but he later stepped back from a full endorsement, leaving a role for floating ice. This hesitant beginning set the tone for the next few decades in North America as its geologists began to debate whether they could see the work of ice sheets or icebergs.

There was a particularly strong tradition of scriptural geology in 19th-century North America. Its practitioners attempted to reconcile their field observations with the Bible and there were often close links with like-minded souls in Britain. *Reliquiae Diluvianae* (1823) was required reading for all at this time. Like their European counterparts, the American geologists of the early 19th century such as Amos Eaton (1776–1842) and Hitchcock himself invoked the power of the deluge to explain valleys and the superficial ('diluvial') deposits of the Quaternary. In many respects, Hitchcock was the William Buckland of New England, and both married women who were gifted illustrators of geological phenomena. Orra White Hitchcock painted superb large format geological illustrations that her husband used in his classes. The Amherst image collection includes a wonderful reproduction of the Adams Mammoth and other ice age beasts including the American Mastodon. There was clearly a good deal of common ground in the 1830s between Hitchcock's geological lectures in Amherst and those of William Buckland in the Old Ashmolean in Oxford.

A Lancashire lad in Connecticut

The beginnings of the ice sheet versus iceberg debate in North America include an intriguing sub-plot from an unlikely source. One Peter Dobson (1784–1878) of Preston in the northwest of England, a master spinner, skilled in the construction of cotton machinery, set sail for Boston in 1809 to set up a cotton mill in Connecticut. Dobson has been hailed as the first person to advocate a role for floating ice in the transport and modification of Quaternary sediments. He departed for the New World in rather inauspicious circumstances after being declared a bankrupt in England. To avoid censure by the authorities before his ship sailed for America (Britain did not want American competition for its own cotton industry) he had to conceal himself, and his plans for a new factory, inside a barrel on the quayside in Liverpool!

Largely self-taught, Dobson was proficient in mathematics, mechanics, and natural history. In 1822, when the foundations were being dug for his new cotton mill in the Talcottville Gorge near Vernon, Connecticut, he made a careful study of the sandstone boulders and gravels that his workmen dug out of the Quaternary deposits. In November 1825, he reported his observations to Benjamin Silliman in Yale, who published his letter in the *American Journal of Science and Arts* the following summer (July 1826):

> I have had occasion to dig up a great number of bowlders of red sandstone and of the conglomerate kind, in erecting a cotton manufactory, and it is not uncommon to find them worn smooth on the underside, as if by their having been dragged over rocks and gravelly earth in one steady direction. On examination, they exhibit scratches and furrows on the abraded part. I think that we cannot account for these appearances unless we call in the aid of ice along with water, and that they have been worn by being suspended and carried in ice over rocks and earth under water.

Dobson sent a more detailed report to Edward Hitchcock in 1838 but he, rather mysteriously, kept it hidden (without reply) for six years before making it public. In this communication, Dobson demonstrated impeccable uniformitarian instincts by suggesting that the drift ice off the Labrador coast might provide a suitable analogue for the process he inferred. These observations were effectively ignored for the best part of two decades until Roderick Murchison saw fit to revive them in London in his Presidential Address to the Geological Society in 1842. Murchison, of course, was dead set against the idea of an ice sheet in Britain (he devoted no less than 17 pages of his address to attack the glacialists) and he seized upon this early American endorsement of his favoured iceberg mechanism. In an attempt to undermine the glacialists, Murchison mischievously declared that Dobson's letter was:

> a short, clear, and modest statement, which, though little more than a page in length, contains the essence of the modified glacial theory, at which we have arrived after so much debate.

> I take leave of the glacial theory in congratulating American science on having the original author of the best glacial theory, though his name had escaped notice, and in recommending to you the terse argument of Peter Dobson, a previous acquaintance with which might have saved volumes of disputations on both sides of the Atlantic.

This was a typical Murchison performance. Here he was, mocking the glacial theory and then attributing the solution to the drift problem, not to Agassiz, Buckland, or Lyell, but to a hitherto unknown cotton manufacturer from Vernon, Connecticut. Of course Murchison should have thanked *Lancashire* science for this singular insight as Dobson was a Preston lad born and bred.

Timothy Conrad and the first signs of glacial ice in America

Another key contribution from this period came from Timothy Abbott Conrad (1803–77), a New York state geologist, sometime poet, and expert on fossil molluscs. Conrad was familiar with the Neuchâtel Discourse and made some very astute early observations about landscape evolution and the Quaternary deposits in the American northeast. These led him to reject both the diluvial and floating ice theories well before Agassiz arrived in America and before the publication of *Études sur les glaciers* in 1840. Conrad noted planed bedrock surfaces and scratches on the hard rocks of western New York State. These, he argued, were produced by a large body of glacial ice that had moved southwards from Canada.

Lyell versus Agassiz

Drift ice was a topic of popular concern beyond geological circles because it formed a major hazard to transatlantic shipping. British and American newspapers regularly published evocative reports of icebergs in the North Atlantic. Lyell himself was so delighted to see sediment-laden drift ice at close quarters when he crossed the Atlantic by steamer in 1845 that he wrote to his sister about it! For Lyell, Murchison, and others, the drift ice of the present was indeed the key to understanding the Quaternary deposits of the past.

As the idea of large-scale continental glaciation began to gain a foothold in North America, it is essential to appreciate that the views of Charles Lyell exerted just as much, if not more, influence in North America as those of Agassiz—even after Agassiz was permanently resident in Boston. Lyell published two books (in 1845 and 1849) documenting his travels and geological observations in North America. It was somewhat ironic therefore that the two men who had presented a united front in London in 1840 should have divided the opinion of geologists quite so profoundly and for so long on both sides of the Atlantic.

If the standing of Lyell extended the useful lifespan of the iceberg theory, it was gradually worn down by a growing body of field evidence from Europe and North America that pointed to the action of glacier ice. Charles Whittlesey (1808–86) recognized moraines and related glacial sediments in Wisconsin in the early 1860s and James Dwight Dana (1813–95), in his seminal *Manual of Geology* (1862), provided a detailed examination of the relative merits of the iceberg and glacier theories in North America and concluded, without equivocation:

> …the glacier theory is alone capable, as first shown by Agassiz, of explaining all the facts.

Writing in *The Popular Science Monthly* of July 1873 (the year that Agassiz died), the State Geologist for Minnesota, Newton Horace Winchell (1839–1914), neatly summarized the state of the debate in America at that time:

> At the present day, but few geologists can be found in this country who do not admit to the reality of the glacial epoch. But, while it is true that but few geologists can be found in this country who do not admit the truth of the glacier theory of Prof. Agassiz, it is also true that a great many, perhaps the majority, also adhere to the iceberg theory of Peter Dobson. The two theories at first came in violent conflict. They diverged at the outset. One required the continent below the ocean, and the transportation of bowlders and other drift by floating ice; the other required it elevated high above the ocean, and the transportation of the drift by ice in the form of continental glaciers.

The continental glacial theory prevailed in North America because it provided a much better explanation for the vast majority of the features recorded in the landscape. The striking regularity and fixed alignment of many features could not be the work of icebergs whose wanderings were governed by winds and ocean currents. The southern limit of the glacial deposits is often marked by pronounced ridges in an otherwise low-relief landscape. These end moraines mark the edge of the former ice sheet and they cannot be formed by floating ice. It took a long time to put all the pieces of evidence together in North America because of the vast scale of the territory to be mapped. Once the patterns of erratic dispersal, large-scale scratching of bedrock, terminal moraines, drumlin fields, and other features were mapped, their systematic arrangement argued strongly against the agency of drifting ice.

Unlike their counterparts in Britain, who were never very far from the sea, geologists working deep in the continental interior of North America found it much easier to dismiss the idea of a great marine submergence. Furthermore, icebergs just did not transport enough sediment to account for the enormous extent and great thickness of the Quaternary deposits. It was also realized that icebergs were just not capable of planing off hard bedrock to create plateau surfaces. Neither were they able to polish, scratch, or cut deep grooves into ancient bedrock. All these features pointed to the action of land-based glacial ice. Slowly, but surely, the reality of vast expanses of glacier ice covering much of Canada and the northern states of the USA became apparent.

The geological survey in Britain and Ireland

In 1855, Roderick Murchison succeeded Henry de la Beche to become Director General of the Geological Survey of Great Britain and Ireland. In the same year a young Scottish geologist named Archibald Geikie (1835–1924) joined the Survey. His younger brother James joined seven years later. In the two decades after 1855 a new breed of professional geologist emerged in Britain as, one by one, the founding fathers of British geology including Buckland, Lyell, Murchison, De la Beche, and Sedgwick, departed the scene. This new generation brought fresh approaches and different ideas. An important figure who bridged these generations was another Scot, this time from Glasgow—Andrew Crombie Ramsay (1814–91), who had joined the survey in 1841.

Ramsey worked mainly in North Wales, the Geikie brothers in the Highlands and Lowlands of Scotland. These men were especially influential in establishing a glacial school in Britain at a time when Lyell's drift ice model still received widespread support. Murchison was an inspirational leader but he liked things done his way. A key legacy of his time at the Survey is the use of the term 'drift', at his insistence, on all of its geological maps. This practice continued for the next 150 years.

Lyell's glacial sea

Some kind of marine submergence was clearly essential to Lyell's drift model to allow melting icebergs to deposit their sediment load across Britain and produce a thick cover of superficial sediment. In 1863, Lyell showed his glacial sea in the form of a map (Figure 19). It is necessary to ignore the modern coastlines to picture Lyell's vision for the geography of northwest Europe during his great submergence. Only the loftiest peaks of Britain and Ireland are dry land—a mainly Celtic archipelago within an ice-infested ocean. One has to imagine fleets of icebergs drifting through narrow straits etching and polishing the exposed bedrock, with their sediment-shod bottoms scraping across submerged bedrock surfaces. In deeper waters the melting bergs released their debris to form thick drift deposits on the sea floor.

Norway and Sweden were obvious potential sources of floating ice because Scandinavian erratics had been discovered at many locations along the east coast of Britain. However, two fundamental weaknesses can be identified in Lyell's

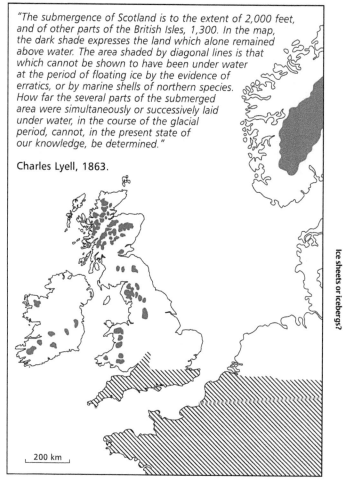

> "The submergence of Scotland is to the extent of 2,000 feet, and of other parts of the British Isles, 1,300. In the map, the dark shade expresses the land which alone remained above water. The area shaded by diagonal lines is that which cannot be shown to have been under water at the period of floating ice by the evidence of erratics, or by marine shells of northern species. How far the several parts of the submerged area were simultaneously or successively laid under water, in the course of the glacial period, cannot, in the present state of our knowledge, be determined."
>
> Charles Lyell, 1863.

200 km

Ice sheets or icebergs?

19. Lyell's glacial sea

palaeogeography. First, his reconstruction leaves a much reduced area of upland landscape that could not sustain glaciers of sufficient size to produce large volumes of icebergs. Second, with sea level so much higher, these mountains were now too low to have a climate that would produce glaciers. Lyell's great marine submergence also became increasingly untenable when it was recognized that the build-up of large masses of ice on the continents must be associated with a significant *fall* in sea level.

Box 4
Early ideas on sea level change

Hutton and Playfair thought that sea level had remained more or less stable over time. It was another Scot, Charles MacLaren, writing in 1842, who first argued that the formation of large ice sheets would result in a fall in sea level as water was taken from the oceans and stored frozen on the land. This insight triggered a new branch of ice age research—sea level change. This topic can get rather complicated because as ice sheets grow, global sea level falls. This is known as *eustatic* sea level change. As ice sheets increase in size, their weight depresses the crust and relative sea level will rise. This is known as *isostatic* sea level change. Thomas Jamieson (1829–1913) first put forward the radical suggestion that the mass of glacial ice might depress the land surface when he observed Pleistocene marine deposits high above modern sea level in the Forth Valley. It is often quite tricky to differentiate between regional-scale isostatic factors and the global-scale eustatic sea level control.

Turning the tide

In Britain and Ireland, as in other parts of the world, the steady accumulation of evidence over many years of field mapping played a fundamental role in gaining wide acceptance for the glacial theory. The Geological Survey mapped moraines, drumlins, glacial lakes, meltwater channels, and other glacial features. By recording many miles of exposures revealed by railway cuttings and the construction of Victorian sewers, the nature and extent of the Quaternary deposits across Britain was established.

Writing in the *Quarterly Review* in 1879, the great naturalist, Alfred Russel Wallace (1823–1913), penned a review essay on the 'Glacial Epoch'. In the opening paragraph he observed:

> ...the glacialists were for some time looked upon as a set of wild enthusiasts, whose facts were not worth examining, and whose theories might be ridiculed or despised. Soon, however, the tide began to turn.

By the late 1870s the tide had indeed turned: glacial geology had become a serious scholarly pursuit with a rapidly growing literature. Charles Lyell died in 1875 and, between the publication of the first (1874) and second (1877) editions of his landmark book *The Great Ice Age*, James Geikie rejected the idea of a glacial marine submergence and became totally convinced of the dominant role of land ice. But geologists such as William Boyd Dawkins, Thomas Bonney (1833–1923), and Joseph Prestwich (1812–96) in Britain, and William Dawson (1820–99) in Canada, clung to the drift ice theory for the lowlands as advocated by Lyell, Darwin, and Murchison. The glacialists had to work extra hard to wear down this opposition. In March 1879, James Geikie made his frustrations very clear in a letter to the wilderness conservation pioneer and Scottish-born American, John Muir (1838–1914):

> I am glad that you found my paper on the glaciation of the Outer Hebrides of interest. I quite agree with you that so many details as I give are not needed to convince anyone who is conversant with the modes of glacial action that these islands [the Outer Hebrides] have been glaciated from the South east. But the 'prejudices' (if I may be allowed to use so hard a word) of English geologists in favour of icebergs are hard to overcome, and one can only hope to do so by piling up the evidence that tells against that most inept and unsatisfactory iceberg hypothesis.

As the Geikie brothers and their colleagues at the Geological Survey continued 'piling up the evidence' into the 1880s, British glacialists received an important boost from North America that would eventually prove fatal for the iceberg theory.

An American abroad in the 1880s

Forty-five years on from the celebrated tour of Scotland by Agassiz and Buckland, another gifted young professor, this time from Philadelphia, began his own glacial tour of Britain and Ireland. This one spanned three summers (1885 to 1887) and it was undertaken, not by horse-drawn stage, but by rail. It passed off without fuss or fanfare, but its findings were to have a profound impact on the glacial debate. Henry Carvill Lewis was born into a wealthy Philadelphia family in 1853. After graduating from the University of Pennsylvania, he joined the State Geological Survey as a volunteer, mapping glacial deposits. He published extensively on mineralogy, on the origin and properties of diamonds, and on glacial geology and ice age environments. He was appointed to the Chair of Geology in Haverford College, Pennsylvania in 1883.

Carvill Lewis loved travelling by train. In the summers of 1885 to 1887, he travelled hundreds of miles in the company of his wife, traversing the landscapes of Britain and Ireland. Through many miles of field walking and by making observations from the window of his carriage, he traced the end moraine complex from the last glacial advance across Ireland, Wales, and England—from Tralee in County Kerry on the Atlantic coast to Flamborough Head on the North Sea. These terminal moraines marked the southern limit of what became known as the Northern Drift. Carvill Lewis argued that all of the drift deposits to the north of this moraine had been deposited, not by icebergs, but by a land-based ice sheet—they were identical in many respects to the glacial sediments he had encountered in Pennsylvania.

When the British Association met in Manchester in September 1887, Carvill Lewis argued that the so-called drift deposits of Britain and Ireland, like those in North America, were the product of land ice and the margins of the ice sheet that produced them could be identified by tracing the end moraines. He also suggested that the glacial deposits at the margins of the Greenland ice sheet were a very good analogue for what he had observed on his tour of the British Isles.

Carvill Lewis also put forward the radical suggestion that the shells at Moel Tryfan and other elevated localities (which provided the most important evidence for the great marine submergence of Britain) were not *in situ*. Building on the earlier suggestions of Thomas Belt (1832–78) and James Croll, he argued that these materials had been dredged from the sea bed by glacial ice and pushed upslope so that 'they afford no testimony to the former subsidence of the land'. Together, his recognition of terminal moraines and the reworking of marine shells undermined the key pillars of Lyell's great marine submergence. This was a crucial step in establishing the primacy of glacial ice over icebergs in the deposition of the drift in Britain. Without the glacial sea, the iceberg theory was sunk.

Whilst Agassiz's findings were hastily reported in *The Scotsman* in October 1840, this episode in the glacial story only made the British press the following year when Henry Carvill Lewis died in his lodgings in Victoria Park in Manchester in July 1888 at the age of just 34. He had been taken ill with blood poisoning (contracted from drinking impure water) during the voyage from New York to Liverpool. The young professor from Philadelphia was buried on 24th July, exactly 51 years to the day after Agassiz had delivered the Neuchâtel Discourse.

So a cotton man from Preston, Lancashire, who moved his livelihood to New England in the first decade of the century, began a debate in 1822 that vexed some of the greatest geologists for over 60 years. Dobson's iceberg mechanism, refined by Lyell and vigorously promoted by Murchison, was finally undermined by a young geologist from Philadelphia who crossed the Atlantic in the other direction in the 1880s. Carvill Lewis was buried in the boulder clay near Bolton, another Lancashire cotton town, just 20 miles south of Preston.

James Geikie's Newcastle address of 1889

By the end of the 1880s, it was the glacial dissenters who formed the eccentric minority. Indeed, in 1887 Archibald Geikie was confident enough to write the statement that opens this book. This shift is clear from the papers of the 1889 meeting of the British Association in Newcastle where his brother James, as President of the Geology Section, delivered the opening address. A year on from the death of Carvill Lewis, the topic of James Geikie's landmark address was the domain of glacial geology:

> Perhaps there is no department of geological enquiry that has given rise to more controversy than that which I have selected for the subject of this address. Hardly a single step in advance has been taken without vehement opposition. But the din of contending sides is not so loud now—the dust of the conflict has to some extent cleared away, and the positions which have been lost or maintained, as the case may be, can be readily discerned. The glacialist who can look back over the last twenty-five years of wordy conflict has every reason to be jubilant and hopeful.

James Geikie went on to point out what he saw as a very British propensity to look to the sea to explain the geological record:

> I have often thought that whilst politically we are happy in having the sea all round us, geologically we should have gained perhaps by its greater distance. At all events we should have been less ready to invoke its assistance to explain every puzzling appearance presented by our glacial accumulation.

The glacial debate formed a central part of the scientific enlightenment of the 19th century. By the close of the century most geologists recognized the impotence of icebergs and accepted that great ice sheets had once formed in now temperate regions during an epoch of Arctic climate. A century on from the discovery of the Adams Mammoth, the ice age and shifting climates had become part of mainstream natural science.

Chapter 7
Glacials, interglacials, and celestial cycles

The imperfection of all our records of the past is too well known to geologists.

Alfred Russel Wallace (1879)

In the period leading up to World War One, there was much debate about whether the ice age involved a single phase of ice sheet growth and freezing climate (the monoglacial theory) or several phases of ice sheet build up and decay separated by warm interglacials (the polyglacial theory). Archibald Geikie was one of the first to recognize evidence for interglacials in Britain when, in the 1870s, he found the remains of warmth-loving plants in peat deposits between two glacial tills. One of the foremost American geologists of this era, Thomas Chrowder Chamberlin (1843–1928), was the first to identify evidence for multiple Quaternary glaciations in North America. The discovery of fossil-rich deposits representing warm interglacial conditions raised fundamental new questions that would shape ice age research for the next century. How many glacials and interglacials were there during the Quaternary? How long did they last? Most crucially, if the geological record showed an ice age of multiple glaciations, a mechanism was needed to explain how the Earth's climate system could shift repeatedly in and out of glacial mode.

The Quaternary Ice Age

In 1914, William Bourke Wright (1876–1939) of the Irish Geological Survey published the first edition of his classic book *The Quaternary Ice Age*. Wright's book, with a gold-engraved mammoth on the cover, presented the first full reconstruction of the British and Irish ice sheet (Figure 20). The margins of this ice sheet were established by mapping end moraines and the limits of glacial deposits. The flow lines, based on striations and the paths of erratics, show the movement of ice radiating from centres of accumulation in Scotland, Wales, and Ireland as well as the influence of Scandinavian ice from the northeast. This landmark in ice age research was the culmination of decades of work by many individuals. It had been quite a journey since the momentous dispatches of Buckland and Agassiz almost 75 years earlier. Interestingly, Wright initially rejected the evidence for interglacials in Britain. When his book was published he was a strong advocate of the view that there had been just one major phase of glaciation during the Quaternary. This monoglacial stance was strongly contested by the Geikie brothers and by the latest research from the Alps.

The Alpine model of Quaternary glaciation

Between 1901 and 1909, Albrecht Penck (1858–1945) and Eduard Brückner (1862–1927) studied the moraines and river terraces in the valleys of southern Germany that drain the Bavarian Alps. They recognized four distinct stages of Quaternary glaciation, naming them, from oldest to youngest, *Gunz*, *Mindel*, *Riss*, and *Würm*, after the rivers in the valleys where the evidence was preserved. The interglacials were associated, not with fossil-bearing sediments, but with phases of erosion as the rivers cut down into the thick deposits of outwash gravels deposited during the preceding glacial stage. Penck and Brückner published a monumental three-volume study in 1909: *Die Alpen im Eiszeitalter* (The Alps in the Ice Age). Their model of four glacial and four interglacial stages became a cornerstone of ice age research. It dominated Quaternary research for the next 60 years.

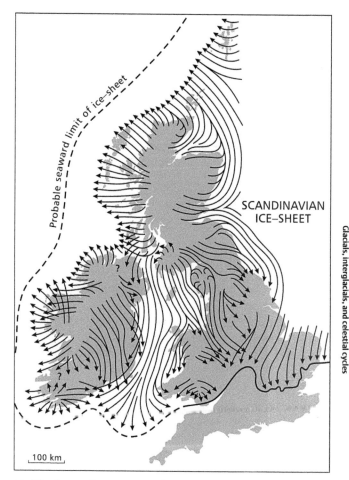

20. **The first modern reconstruction of the British and Irish ice sheet**

The astronomical theory of the ice ages

Penck and Brückner had provided the first geological template for
the Quaternary ice age. It was a Serbian engineer, Milutin
Milankovitch (1879–1958), who made it his life's work to explain

these ice age rhythms. Milankovitch was convinced the answer was to be found by examining astronomical factors—the changing relationship between the Earth and the Sun and how this influenced the Earth's seasonal energy budget over geological timescales. Milankovitch built on the pioneering work of James Croll, the self-taught Scottish scientist who, between about 1860 and 1880, had developed his own theory of ice age climate change based on variations in insolation. In different ways, both men overcame formidable obstacles to make seminal contributions to ice age research. Both were decades ahead of their time.

Box 5
Insolation

This is the solar radiation received over a given surface area over a unit of time. It can be expressed in various ways including megajoules of energy per square metre (MJ/m^2) or watts per square metre (Wm^2). You will be familiar with these units if you have solar panels on your house. It is a critical parameter for the viability of snow patches and glaciers because insolation varies significantly across latitudes, between seasons, and over geological timescales.

Earth seasons

The Earth is tilted with respect to the plane of its orbit around the Sun and this tilt produces our seasons (Figure 21). The hemisphere that is angled towards the Sun is in summer whilst the other hemisphere is in winter. Without this tilt we would not have seasonal contrasts because all parts of the atmosphere would receive the same amount of solar radiation throughout the year and there would be no difference in the length of the days between hemispheres. Today, summer is warmer than winter because: (1) the days are longer (>12 hours) so the Sun provides solar energy for longer; and (2) the Sun is higher in the sky so that the land surface receives more solar energy per unit area. Winter days are shorter (<12 hours) and the Sun is lower in the sky, so insolation is also lower. This may seem like rather basic stuff, but if this familiar pattern changes—even by quite small amounts—there can be far-reaching effects on the climate system.

Eccentricity, obliquity, and precession

As the Earth rotates about its axis travelling through space in its orbit around the Sun, there are three components that change over time in elegant cycles that are entirely predictable. These are known as eccentricity, precession, and obliquity or 'stretch, wobble, and roll' (Figure 21). These orbital perturbations are caused by the gravitational pull of the other planets in our Solar System, especially Jupiter. Milankovitch calculated how each of

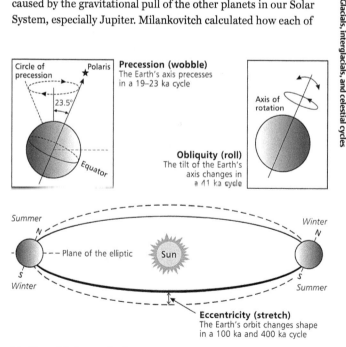

Precession (wobble)
The Earth's axis precesses in a 19–23 ka cycle

Obliquity (roll)
The tilt of the Earth's axis changes in a 41 ka cycle

Eccentricity (stretch)
The Earth's orbit changes shape in a 100 ka and 400 ka cycle

Circle of precession — Polaris
23.5°
Equator

Axis of rotation

Summer N — Winter N
Plane of the ecliptic — Sun
Winter S — Summer S

21. The orbital perturbations

these orbital cycles influenced the amount of solar radiation received at different latitudes over time. These are known as Milankovitch Cycles or Croll–Milankovitch Cycles to reflect the important contribution made by both men.

Orbital eccentricity

The shape of the Earth's orbit around the Sun is not constant. It changes from an almost circular orbit to one that is mildly elliptical (a slightly stretched circle) (Figure 21). This orbital eccentricity operates over a 400,000- and 100,000-year cycle. With an elliptical orbit the Earth is slightly closer to the Sun for part of the year and further away for the rest. Today, when the Earth is closest to the Sun, the distance between them is about 147.1 million km. This is known as *perihelion* and it takes place around 3rd January. When the Earth is at its furthest point from the Sun (*aphelion*) around 4th July, the distance increases to 152.1 million km. The northern hemisphere is still cold in winter at perihelion because of the tilt of the Earth. Likewise the southern hemisphere is warm in its summer at aphelion because of the tilt. Changes in eccentricity have a relatively minor influence on the total amount of solar radiation reaching the Earth, but they are important for the climate system because they modulate the influence of the precession cycle (see below). When eccentricity is high, for example, axial precession has a greater impact on seasonality.

Obliquity

The Earth is currently tilted at an angle of 23.4° to the plane of its orbit around the Sun. Astronomers refer to this axial tilt as obliquity. This angle is not fixed. It rolls back and forth over a 41,000-year cycle from a tilt of 22.1° to 24.5° and back again (Figure 21). Even small changes in tilt can modify the strength of the seasons. With a greater angle of tilt, for example, we can have hotter summers and colder winters. With less tilt the opposite would be the case. Cooler, reduced insolation summers are thought to be a key factor in the initiation of ice sheet growth in the middle and high latitudes because they allow more snow to survive the summer melt season. Slightly warmer winters may also favour ice sheet build-up as greater evaporation from a warmer ocean will increase snowfall over the centres of ice sheet growth.

Precession

The Earth's axis of rotation is not fixed. It wobbles like a spinning top slowing down. This wobble traces a circle on the celestial sphere (Figure 21). At present the Earth's rotational axis points toward Polaris (the current northern pole star) but in 11,000 years it will point towards another star, Vega. This slow circling motion is known as axial precession and it has important impacts on the Earth's climate by causing the solstices and equinoxes to move around the Earth's orbit. In other words, the seasons shift over time. Precession operates over a 19,000- and 23,000-year cycle. This cycle is often referred to as the *Precession of the Equinoxes*. Today perihelion is in January but in about 11,000 years' time it will be in July, and northern hemisphere summers will be warmer than they are now.

James Croll (1821–90)

James Croll was the leading 19th-century advocate of an orbitally driven theory of ice age climate change. He was inspired by the French mathematician, Joseph Adhémar (1797–1862), who was the first to propose a link between ice age climates and astronomical forces.

Adhémar also advanced the radical notion that ice ages alternated between the hemispheres. He argued that the southern hemisphere was currently in an ice age whilst interglacial conditions prevailed in the northern hemisphere. Inspired by Agassiz's ice age, he published these ideas in 1842 in his book *Revolutions of the Sea*.

Croll's career followed an unusual path. He was variously employed as an apprentice millwright, tea shop owner, insurance agent, and manager of a temperance hotel before becoming a caretaker in 1859 at the Andersonian College and Museum in Glasgow (that would later become the University of Strathclyde). It was here that he gained access to the books and journals he needed to work on his astronomical theory. Croll corresponded with Lyell and Darwin, and later took a clerical post at the Scottish Geological Survey where he was encouraged by the Geikie brothers and Andrew Ramsay. Croll's thinking was decades ahead of its time in several key ways. He considered the role of ocean currents and ice albedo effects, for example, alongside his general model of climate change—although he was wrong to suggest that ice ages alternated between the southern and northern hemispheres. Charles Lyell was not at all convinced by the suggestion of extra-terrestrial drivers of ice age climates, but he was eventually persuaded of their potential importance by the pioneering work of James Croll. The Geological Society awarded Croll its highest honour, the Wollaston medal, in 1872.

Box 6
Albedo and feedback effects

The albedo of a surface is a measure of its ability to reflect solar energy. Darker surfaces tend to absorb most of the incoming solar energy and have low albedos. The albedo of the ocean surface in high latitudes is commonly about 10 per cent—in other words, it absorbs 90 per cent of the incoming solar radiation. In contrast, snow, glacial ice, and sea ice have much higher albedos and can reflect between 50 and 90 per cent of incoming solar energy back into the atmosphere. The elevated albedos of bright frozen surfaces are a key feature of the polar radiation budget.

Albedo feedback loops are important over a range of spatial and temporal scales. A cooling climate will increase snow cover on land and the extent of sea ice in the oceans. These high albedo surfaces will then reflect more solar radiation to intensify and sustain the cooling trend, resulting in even more snow and sea ice. This positive feedback can play a major role in the expansion of snow and ice cover and in the initiation of a glacial phase. Such positive feedbacks can also work in reverse when a warming phase melts ice and snow to reveal dark and low albedo surfaces such as peaty soil or bedrock.

Milankovitch and insolation

After beginning his great project at Belgrade University in 1912, Milankovitch's career was beset with turmoil as he became entangled with some of the major conflicts of Balkan history. The University library was sacked during World War One and, in 1914, whilst on honeymoon, Milankovitch was arrested by the Austro-Hungarian military police. He became a prisoner of war, interned in Budapest for four years. But he kept thinking about ice age climates and carried on doing the maths. Using the rules of celestial mechanics, he did all of the complex mathematics to calculate very precisely how the changing geometrical relationships between the Earth and the Sun affected the receipt of solar energy at different latitudes. It was a monumental task. No calculator, no spreadsheets—just a pencil and paper, a book of logarithms, a slide rule, and some very tricky equations. It took Milankovitch the best part of two decades to complete the calculations, compile the endless tables of data, and perfect his theory.

His most famous work: *Canon of Insolation and the Ice-Age Problem* was published by the Royal Serbian Academy in 1941. The manuscript was delivered to the printers in early April just four days before Belgrade was devastated by German bombing. As the city burned, the national library and its entire collection was completely destroyed. The press that had just printed the Milankovitch opus was also gutted. Quite miraculously, almost all the pages survived in the warehouse and those that were damaged were reprinted. His life's work, published in German, was very nearly consumed by the flames from Luftwaffe bombs.

The impact on climate

Milankovitch concluded that the key driver of ice age climate change was the amount of solar radiation received in the summer time in the northern hemisphere. In the 1870s, James Croll had taken the opposite view, believing that the severity of winters was the key factor. Summer temperatures, however, are critical because they determine how much snow and ice is lost through melting. Perhaps the greatest achievement of Milankovitch was the way he explored the *combined* influence of the three orbital cycles for the last 600,000 years and theorized about the optimal orbital configurations that would produce the maximum change in climate. His calculations suggested that latitudinal shifts in insolation could render the summer time at 65°N of today (Iceland) more like summer at 77°N (Spitzbergen). In crude terms, this means the Arctic Circle shifting several hundred kilometres south to the high temperate zone so that ice sheets could form there and extend down to sea level.

The Milankovitch theory was supported by the famous German climatologist Vladimir Köppen (1846–1940) and by Alfred Wegener (1880–1930), the pioneer of continental drift. They attempted to correlate the orbital cycles with the Penck and Brückner scheme. But the astronomical theory was also very heavily criticized by many geologists who needed to see hard evidence. The model lacked solid empirical support from the geological record. In 1936, Penck himself told Milankovitch that he thought his ice age theory was hogwash!

There is no doubt that the farsighted ideas of Croll and especially Milankovitch produced a major leap forward in the quest to identify the cause of the ice ages. Despite its elegance, however, the astronomical theory fell out of favour because it could not be subjected to rigorous testing. There was no independent timescale for the Quaternary—the fragmentary geological record of the glaciated regions was just not up to the task. The moraines in the Alps were not dated and there was no way of working out the length of glacials and interglacials or the size of any gaps in the known sedimentary records. A new approach was needed.

Chapter 8
Deep ocean sediments and dating the past

The sea must know more than any of us.

Carl Sandburg (1918)

Away from the continents, in the low energy setting of the deep ocean floor, fine sediments dominated by the shells of tiny marine organisms have accumulated over millions of years to form a thick veneer of soft mud. Marine sediments are especially useful because they form an unbroken record of environmental change that spans the entire Quaternary. Geologists had long speculated that these deposits might tell the full story of the Quaternary ice age. So when the Swedish oceanographer, Börje Kullenberg (1906–91), designed a piston corer in 1947 that could achieve deep penetration into the soft sediments on the ocean floor and recover long undisturbed cores, it heralded a new era in ice age research. In 1968, the Deep Sea Drilling Project began to collect hundreds of long sediment cores. Two aspects of the marine sediment record proved to be particularly instructive for reconstructing the history of the terrestrial ice sheets: the isotope chemistry of foram shells and the incidence of ice-rafted debris (IRD) (Figure 22).

Forams and their shells

Forams are tiny unicellular creatures found at all depths in polar, temperate, and tropical marine environments. Planktonic forams

22. a) The fossil shell of the planktonic foram *Globigerinoides ruber*
(b) Sand-sized ice-rafted debris with a small number of forams from a
North Atlantic sediment core

inhabit the upper part of the water column that is penetrated by
sunlight whilst benthic forams live close to the sea floor in the
darkness of the deep ocean. Many species build pinhead-sized
shells of calcium carbonate ($CaCO_3$) from elements extracted from
seawater—these shells record key properties of the waters in which
they live. When forams die and sink to the sea bed they contribute
to a neatly stacked sedimentary record of changing ocean
conditions from the present day to the deep past. Figure 22a shows
a fossil foram shell viewed under a scanning electron microscope.
In living specimens thousands of thin organic filaments or
rhizopodia extend through the holes in the shell to help the foram
move and feed. Some rather ingenious approaches have been
employed to prise information from foram shells to illuminate
aspects of the ice age past. One of these is oxygen isotope analysis.

The marine oxygen isotope record

Quaternary research was transformed in 1973 when the record
shown in Figure 23 was published by Nick Shackleton and Neil

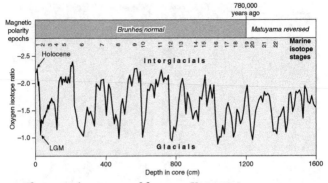

23. The oxygen isotope record from core V28-238

Opdyke. This is the oxygen isotope record from marine sediment core V28-238. It is perhaps the most famous plot in ice age research since it revolutionized our understanding of Quaternary ice sheet dynamics and global environmental change. It is therefore worth thinking about how it was derived, what it represents, and why it had profound implications for all of the natural sciences.

Oxygen has three stable isotopes: ^{16}O, ^{17}O, and ^{18}O. The lightest (^{16}O) and heaviest (^{18}O) are widely utilized in the study of past environments. Due to a process known as isotopic fractionation, when evaporation takes places from a body of water (H_2O), more of the isotopically lighter (^{16}O) water molecules are removed from the water surface relative to the isotopically heavy (^{18}O) water molecules because less energy is required to evaporate the lighter isotope. Fractionation operates at all scales: from the surface of the smallest pond to the largest ocean. In the global hydrological cycle, therefore, the ratio of $^{18}O/^{16}O$ in water vapour is less than the $^{18}O/^{16}O$ ratio of the seawater from which it is derived. The differences are very small, just a few parts per thousand, but they can be measured to high precision.

Over many thousands of years, as ice sheets build up during the course of a glacial cycle, enormous volumes of water are abstracted from the oceans and stored on the continents as glacial ice. The fractionation process creates ice sheets that are isotopically 'light' because they are fed by snow formed from atmospheric water vapour. As the ice sheets on land increase in size, the oceans become increasingly isotopically 'heavy'. When ice sheets melt at the end of a glacial stage, the isotopically light meltwater is returned to the oceans. Thus, over glacial–interglacial cycles, the oxygen isotope ratio of ocean water shifts in response to the waxing and waning of the continental ice sheets. In order to exploit this process to explore the ice age past, we need a long record of changes in the oxygen isotope ratio of ocean water. This is provided by the fossil foram shells recovered from marine sediment cores. We can measure the ratio of $^{18}O/^{16}O$ in these shells using a mass spectrometer.

Figure 23 shows the changing isotopic composition of the oceans for the past million years as measured in forams from core V28-238. The record is divided into Marine Isotope Stages (MIS). The deepest troughs in the curve (even numbers) are glacial maxima with high global ice volume. The highest peaks (odd numbers) represent interglacials with much reduced ice volume. MIS 2, for example, is the global Last Glacial Maximum (LGM) (c.22,000 years ago). MIS 1 is the present (Holocene) interglacial. Note that MIS 3 saw a reduction in global ice volume but full

Box 7

V28-238: This code identifies the 238th sediment core collected on the 28th cruise of the research vessel *Vema*. This core (1,600 cm long) was collected just north of the Equator in the Pacific Ocean in a water depth of 3,120 m. *Vema*, originally built as a luxury iron-hulled schooner (*Hussar*) in 1923 by millionaire New York stockbroker Edward Hutton (1875–1962), was refitted to become the main research vessel of the Lamont-Doherty Geological Observatory of Columbia University. Her crews collected thousands of sediment cores from the ocean floor.

interglacial conditions were not attained and MIS 5 is therefore known as the Last Interglacial.

Ocean temperature or terrestrial ice volume?

There are two principal controls on the oxygen isotope ratio of modern marine foram shells: (1) the isotopic composition of the ocean water in which they live; and (2) the temperature of that water. There has been much debate about the contribution of these two factors to the Quaternary isotope record. In the 1940s, at the University of Chicago, Nobel Laureate Harold Urey (1893–1981) was the first to suggest that oxygen isotopes could be used to reconstruct ancient temperatures. Urey encouraged a young colleague, Cesare Emiliani (1922–1995), to apply this principle to the study of the ice age oceans. Emiliani believed that the isotopic fluctuations he observed in planktonic forams in marine sediments were primarily a function of changing ocean *temperature* during glacial and interglacial cycles. He published a classic paper on this work in 1955.

During his PhD research in the mid-1960s Shackleton noticed a fundamental error in Emiliani's work. Shackleton argued that changes in terrestrial ice volume and *not* ocean temperature were the dominant influence on the isotopic variations detected in the foram record. He was able to demonstrate this by analysing benthic forams from the last glacial period. Benthic forams secrete their shells in the dark unchanging environment of the deep ocean where Quaternary water temperatures have always been close to zero. This environment is effectively insulated from the glacial–interglacial shifts in temperature we see on land and in ocean surface waters. This new approach took temperature out of the equation. Shackleton showed that the oxygen isotope record from benthic forams displayed the same fluctuations that Emiliani had observed in planktonic species. The benthic signal must therefore have been controlled by changes in the volume of ice stored on the continents. This was a stunning breakthrough (Figure 24).

24. Nick Shackleton in Cambridge in 1972

Dating the marine record

Using his long isotope records Emiliani attempted to revive the
Milankovitch theory, but he was hampered by poor dating control.
Developing a reliable timescale for the long marine records was
problematic throughout the 1950s and 1960s. Radiocarbon dating

had been developed by another Nobel Laureate in the Chicago group, Willard Libby (1908–80), just after the Second World War. This method could be used to date carbonate shells in the top section of cores back to about 50,000 years, but even this was only a small fraction of Quaternary time. Without a reliable timescale it was not possible to establish the tempo of change or the length of glacials and interglacials.

Box 8
Radiocarbon dating

Willard Libby made the first calculation of the half-life of radiocarbon and recognized its potential as a geological dating tool. The unstable isotope of carbon (^{14}C) is present in tiny quantities in all plants and animals. It is produced in the upper atmosphere by cosmic rays and enters the food chain with stable carbon (^{12}C) via the carbon cycle. All living organisms take up ^{14}C. When an organism dies, its radiocarbon stock is no longer replenished. By measuring the residual radiocarbon in fossil material, such as wood, charcoal, peat, seeds, bone, or foram shells, it is possible to estimate when the organism died using the half-life of $5,570 \pm 30$ years. In the early 1960s, samples of skin and fatty tissues taken from the Adams Mammoth yielded ages of $34,450 \pm 2,500$ and $35,800 \pm 1,200$ years BP (Before Present). By convention, the year 1950 is deemed to be the 'present' for radiocarbon dates.

Specialist equipment is needed to measure radiocarbon—for every trillion (10^{12}) atoms of stable carbon in a living organism, there is just one atom of ^{14}C. In theory, the method can provide ages for samples up to 50,000 years old, but beyond six half-lives this involves the accurate measurement of tiny amounts of residual radiocarbon. This can be done, but such very old samples can be susceptible to contamination by more recent radiocarbon and this yields ages that are too young.

Since Libby pioneered this method there have been important advances in sample preparation and in the way radiocarbon is measured. Also, because radiocarbon production in the upper atmosphere is not constant, all dates must be calibrated because radiocarbon years are not equivalent to calendar years. Despite these difficulties, it remains the most widely applied dating method in ice age research even though it can only be used for the Holocene and the second half of the last glacial stage. Libby was awarded the Nobel Prize for Chemistry in 1960.

Shackleton and Opdyke employed a different approach to date their isotope record using reversals in the Earth's magnetic field. Opdyke made systematic down-core assessments of magnetic polarity and located a reversal in V28-238 at a depth of 1,200 cm in MIS 19 (Figure 23). Because the V28-238 record does not contain any obvious breaks in sedimentation, he could be confident that this was the Brunhes-Matuyama reversal—the last time the Earth's magnetic field flipped—780,000 years ago. From this fixed point ages could be interpolated for each level in the core by assuming, quite reasonably, that the sediments had accumulated at a uniform rate. For the first time, reliable ages could be assigned to the glacial and interglacial stages of the Quaternary.

Box 9
Dating using magnetic reversals

Today a compass needle points toward magnetic north. This has not always been the case because magnetic reversals—when magnetic north becomes magnetic south or *vice versa*—are an intrinsic feature of our planet's magnetic field over geological timescales. The last full reversal took place 780,000 years ago—this is known as the Brunhes-Matuyama reversal. Prior to that reversal, the needle of a magnetic compass would point to magnetic south. All sediments deposited in the ocean include a component of minerals that retain information on the Earth's polarity at the time of deposition. We can measure this palaeomagnetic signal in the laboratory to establish if the sediments were deposited during a period of normal polarity (like today) or reversed polarity. Periods of normal polarity are shown in black and periods of reversed polarity in white (Figure 23). A magnetic reversal does not mean—as a student once excitedly asked this author after a lecture—that the Earth itself flips through 180°! It is the magnetic *field* that flips. The causes of reversals are not fully understood, they are probably triggered by the shifting motions of the hot liquid metals that surround the Earth's core.

Implications: global ice sheet dynamics

The oxygen isotope record from V28-238 has been described as the 'Rosetta Stone' of the ice ages. It shows that for most of the last 1 million years, large ice sheets were present in the middle latitudes of the northern hemisphere and sea levels were lower than today. Indeed, 'average conditions' for the Quaternary Period involve much more ice than present. The interglacial peaks—such as the present Holocene interglacial, with its ice volume minima and high sea level—are the exception rather than the norm. The sea level maximum of the Last Interglacial (MIS 5) is higher than today. It also shows that cold glacial stages (*c.*80,000 years duration) are much longer than interglacials (*c.*15,000 years). Ice volume in Antarctica did increase during glacial stages, but the Quaternary marine isotope record is dominated by the much larger changes in ice volume that took place in the northern hemisphere.

Box 10
CLIMAP

A good deal of the ice age research in the marine realm was made possible because of an international collaboration known as CLIMAP (Climate: Long range Investigation, Mapping, and Prediction) established in 1971 by James Hays of Lamont. It was funded by the United States National Science Foundation. John Imbrie of Brown University, Nick Shackleton, Neil Opdyke, and others were prominent members of this team. A key outcome was the reconstruction of the oceans and land surface of the Earth at the LGM.

The steep transitions from even to odd numbered stages indicate that the large ice sheets decayed very rapidly (Figure 23). These rapid deglaciations are known as Terminations. Indeed, planktonic forams in marine cores from the Gulf of Mexico record a distinctive influx of isotopically light water between 20,000 and 15,000 years ago because large meltwater floods travelled some 2,000 km down the Mississippi Valley from the disintegrating Laurentide Ice Sheet. The isotope stages are considered to be globally synchronous with a resolution that is better than 1,000 years—this is the best estimate for the mixing time for any regional changes to be transmitted throughout the global ocean.

Box 11
Rapid deglaciation

Deglaciations are rapid because positive feedbacks speed up both the warming trend and ice sheet decay. As Milankovitch-induced changes in summer insolation warmed the oceans at the end of glacial stages, the oceans released CO_2 into the atmosphere. This strengthened the greenhouse effect and reinforced the warming trend. Rising sea levels further undermined the ice sheet margins.

This process accelerated melting and hastened the sea level rise. Not only do warmer oceans produce greater volumes of atmospheric water vapour (another potent greenhouse gas), they also reduce the extent of sea ice. As both glacial ice and sea ice diminish, the amount of solar energy reflected back into space also falls because they reveal surfaces with much lower albedos. Internal feedbacks combine to amplify warming.

Figure 24 shows Nick Shackleton in his laboratory in 1972 surveying the oxygen isotope data from core V28-238, holding a test tube of precious forams. When the analysis for V28-238 was complete, Shackleton stuck the long sheet of graph paper around the wall of his living room at home in Cambridge and threw a party. He was fully aware that this record signposted a new direction for ice age research. Many of the fundamental questions about the climate system and ice sheet dynamics could now be tackled by studying the sediments on the ocean floor. In 1976, Shackleton and Opdyke published the record from sister core V28-239 which spanned the entire Quaternary. The ice age now had a template showing the *complete* record of glacials and interglacials.

Oxygen isotope analysis quickly became established as a powerful geological tool for correlating marine records around the world. The records from V28-238 and V28-239 were soon replicated in all the oceans. Shackleton and Opdyke had produced the first truly global, well dated, and continuous record of glacial events on the continents and, by inference, sea level change, for the entire

Box 12
Nicholas John Shackleton

Born in 1937, a century on from the Neuchâtel Discourse, the son of an eminent geologist and a distant relative of Ernest Shackleton of the Antarctic, it would seem that Nick Shackleton was destined to work on the ice sheets of the Quaternary. After graduating in Physics in 1961, he was encouraged by the botanist Sir Harry Godwin (1901–85) to work on isotopes in marine sediments for his PhD. Although he spent his entire career in Cambridge, Shackleton was a great collaborator internationally and became a key member of the CLIMAP group of mostly American geoscientists who pooled their expertise to explore the nature of ice age Earth. Shackleton received a bucketful of the most prestigious awards in recognition of the significance and excellence of his work. He was knighted in 1998 and, in 2010, when the Royal Society celebrated its 350th anniversary, he was chosen with ten notable members of the Fellowship (including Isaac Newton, Benjamin Franklin, Ernest Rutherford, and Alfred Russel Wallace) to feature on a special issue of Royal Mail stamps. Shackleton's other passion was music. He was an accomplished clarinettist and an internationally renowned collector and scholar of that instrument. He invariably used the proceeds of his geological prizes to amass the world's largest collection of rare woodwind instruments. He even lectured on acoustics in the Department of Music in Cambridge. Shackleton's many geological achievements would not have been possible without the selfless support of his good friend and laboratory manager *par excellence* Mike Hall. It was he who ensured that the Cambridge mass spectrometers carried out more isotope analyses on forams than any other laboratory.

Quaternary. The Penck and Brückner model of four glacials and four interglacials was now obsolete.

Milankovitch revived: pacemaker of the ice ages

In the late 1960s, a new record of Quaternary sea level change was produced by dating the uplifted coral reefs of Barbados. Because the reefs lay beyond the range of radiocarbon, they were dated using the uranium-thorium decay series. Sea level high stands were identified at 230,000, 170,000, 125,000, 105,000, and 82,000 years ago: the intervals between were times of low global sea level associated with continental glaciation. Significantly, the ages of the high stands coincided with times of maximum solar radiation in the northern hemisphere. These findings sparked a renewed interest in the astronomical theory.

Shackleton's saw-toothed isotope curve also hinted at some degree of cyclicity: the isotopic ratios of the glacial maxima were very similar and they appeared to be evenly spaced (Figure 23). The ice volume record presented a new opportunity to revisit the orbital theory of ice age climate variability and to test the new sea level model from Barbados. To this end, Shackleton collaborated with James Hays (who led the CLIMAP team) and John Imbrie, who had pioneered the use of computers and a statistical tool called spectral analysis to explore trends in long geological datasets. Shackleton's expertise in acoustics and waves also served him well in this analysis.

Spectral analysis of two long oxygen isotope records from the southern Indian Ocean revealed four dominant peaks that matched closely with the cycles computed by Milutin Milankovitch half a century earlier (Figure 25). Hays, Imbrie, and Shackleton had confirmed that long-term changes in the Earth's climate and geological processes were paced by changes in orbital geometry. This has been described as one of the most important geological discoveries ever made. Of course, it also represented an emphatic vindication of the Serbian engineer's life's work. The astronomical theory wasn't hogwash after all.

This breakthrough also created a new problem. Milankovitch Cycles actually generate quite modest changes in insolation, yet the climate shifts of the Quaternary are evidently rather more dramatic. The processes and feedbacks involved in shifting the Earth from glacial to interglacial conditions and back again were yet to be established.

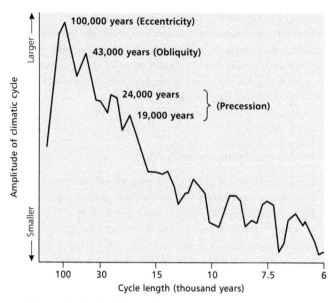

25. **Spectral analysis of the marine oxygen isotope record**

Shackleton and others went on to produce more detailed oxygen isotope records that extended deep into pre-Quaternary time. In the first decade of the 20th century, Penck and Brückner had proposed four glacial–interglacial cycles for the Quaternary when many were making the case for just one. Figure 26a shows a 2.5-million-year record (with a chronology provided by several magnetic reversals) showing no fewer than 50 such cycles over this period back to MIS 104—it is a quite astonishing demonstration of our planet's propensity for change.

This amount of Quaternary climate variability has profound implications for all of the natural sciences. Not only did geologists have to rethink the speed of ice sheet build up and decay (not to mention sea level change), biologists now had to contemplate an order of magnitude shift in the number of times that ecosystems reorganized themselves during the Quaternary ice age. Think back to the vegetation transects across Europe shown in Figure 4. The new complexity from the marine records also challenged geologists who worked on the continents to fill in the gaps and to produce geological records of comparable complexity with much more robust dating frameworks.

Box 13
Quaternary dating methods

Physics has presented geology with many ways of measuring deep time. Several methods are capable of providing reliable age frameworks for deposits that are older than the range of radiocarbon (>50,000 years). Three of the most important are outlined below. There are two key considerations in the choice of dating method: (1) the time range of the technique; and (2) the composition of sample material.

Uranium-thorium dating

This method is based on the measurement of both the parent (uranium-234) and daughter (thorium-230) isotopes using high precision mass spectrometry. It has a range of about 500,000 years and can be used to date carbonate materials such as corals, stalactites, and bones because they retain both uranium and thorium. With a half-life of 245,000 years, ^{234}U decays to ^{230}Th, which has a half-life of 75,000 years. Over thousands of years, as ^{230}Th builds up in a carbonate sample, it provides a precise measure of age. Under optimum sample conditions its range is an order of magnitude greater than that of radiocarbon dating.

Luminescence dating

All geological deposits are exposed to low level natural background radiation. Grains of quartz and feldspar behave as dosimeters because they absorb this ionizing radiation and store trapped electrons in their crystal lattices (traps). This is a time-dependent process that can be used to establish the length of time since burial. Exposure to sunlight empties these traps and resets the geological clock. When grains of quartz or feldspar are exposed to a light or heat source in the laboratory they emit light; this signal is known as *luminescence*. By measuring the strength of this signal and establishing how much annual background radiation the sediments currently receive, it is possible to calculate the length of time since burial. If the stimulus in the laboratory is a light source, the method is called optically stimulated luminescence (OSL). It can be applied over a very wide time range—from very young deposits (c.100 years) to sediments more than a million years old. OSL dating is particularly useful for dating aeolian sediments and fluvial sands deposited on glacial meltwater plains. Such deposits are very often unsuitable for radiocarbon dating because they contain little or no organic material. OSL has also been widely employed to date dune fields and flood deposits in deserts.

Cosmogenic isotope dating

This is a relatively new dating method that utilizes another time-dependent geological process: the build-up of cosmogenic nuclides in rocks and soils exposed at the Earth's surface. These nuclides are created when high energy particles bombard rocks and sediments. Over time, cosmogenic isotopes such as beryllium-10, aluminium-26, and chlorine-36 accumulate in the upper metre or so of the Earth's surface. The duration of exposure can be determined by measuring their concentration. This method has been widely applied in glaciated mountains to date large boulders on moraines. It has a range of about 200,000 years. Many of the moraines and erratic boulders in the Alps that were first mapped in the 19th century have now been dated using cosmogenic isotopes. It has also been used to date bedrock surfaces to establish the timing of ice sheet retreat in Greenland and Antarctica.

The Mid-Pleistocene Revolution

Another key discovery in the long marine records is a marked shift in the *amplitude* of climate change around 900,000 years ago when the duration of glacial–interglacial cycles increased from 41,000 years (paced by obliquity) to 100,000 years (paced by eccentricity). This shift in the length of glacial–interglacial cycles has been termed the Mid-Pleistocene Revolution (MPR) (Figure 26a). The MPR heralded an era of more intensive glaciation when the size of the ice sheets in the northern hemisphere increased significantly. The cause of this striking tempo shift has generated much debate because eccentricity has the weakest influence on insolation. This has been termed the 100,000 year problem. This mismatch between stimulus and outcome suggests there must be internal mechanisms that amplify the response of the climate system. Some researchers have argued that the ice sheets became thicker and perhaps less responsive to the obliquity and precession cycles that previously enhanced summer melting. Glacial stages therefore became longer and terminations only took place when these cycles were strengthened by the eccentricity cycle. It is also likely that other internal mechanisms such as albedo feedbacks and atmospheric greenhouse gas concentrations were important in this remarkable shift in the pacing of glacial cycles.

Ice-rafted debris and Heinrich Events

Where glaciers reach the coast, calving bays produce sediment-laden icebergs. The sediment they transport is known as IRD. In the mid-1980s a young German marine geologist named Hartmut Heinrich found six distinctive spikes in IRD in the North Atlantic marine record for the last glacial stage (Figure 26c). Just a few centimetres in thickness, the Heinrich layers (H1 to H6) can be traced in sediment cores over large sectors of the North Atlantic. They become thinner moving south and east from the Labrador Sea. The IRD is angular, predominantly sand-sized sediment with

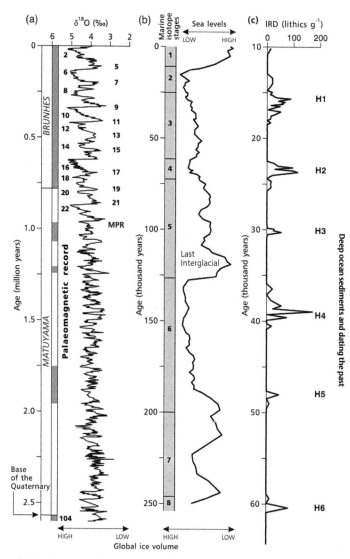

26. Marine oxygen isotope records for (a) the last 2.5 million years and (b) the last 250,000 years; (c) Heinrich Events in the North Atlantic during the last glacial. Note the different timescales for each record

a mix of rock types from the glaciated interior of North America (Figure 22b). Each particle is a tiny erratic that has been transported in glacial ice and set afloat from a calving bay. Heinrich layers form a dramatic record of terrestrial ice sheet instability. Each IRD layer was deposited by an armada of melting icebergs that calved from the Laurentide Ice Sheet and travelled down the Hudson Strait—they represent truly colossal discharges of glacial ice from land to sea. These are known as Heinrich Events.

Mechanisms both internal and external to the Laurentide Ice Sheet have been put forward to account for Heinrich Events. It has been suggested that once the ice sheet reached a critical size, it became unstable and shed mass catastrophically via large-scale sliding and calving. This is the so-called *'binge–purge'* model. Rises in sea level and earthquakes triggered by crustal rebound have also been implicated in undermining the margins of the ice sheet. The identification of Heinrich Events shattered the long held notion that the large ice sheets were fairly stable throughout a glacial period. Indeed, the globally averaged marine isotope curve from tropical latitudes does not show any dramatic change between 60,000 and 10,000 years ago (Figure 26b).

The vast armadas of drifting ice chilled the surface of the Atlantic and checked the influence of the Gulf Stream. Melting icebergs created a layer of isotopically light, low salinity surface water that is clearly recorded in the few planktonic forams found within the Heinrich layers (Figure 22b). Heinrich Events were rather short-lived—lasting for about 750 years. They do *not* represent a revival of the Lyell drift ice hypothesis—they were associated with the collapse of mid-latitude ice sheets and very severe Arctic climates. They produced the coldest and driest conditions in Europe during the last glacial stage. The reduction in evaporation from the cold Atlantic produced widespread droughts that led to collapses in tree populations in southern Europe and dramatic falls in the level of the Dead Sea.

The Cenozoic glacial epoch: the last great cooling

Marine sediments also provide an extended record of global change that has allowed us to gain a deeper understanding of Earth's steady shift from greenhouse to icehouse over the past 55 million years. At the end of the Cretaceous, around 65 million years ago (Ma), lush forests thrived in the Polar Regions and ocean temperatures were much warmer than today. This warm phase continued for the next 10 million years, peaking during the Eocene thermal maximum (Figure 27). From that time onwards, however, Earth's climate began a steady cooling that saw the initiation of widespread glacial conditions, first in Antarctica between 40 and 30 Ma, in Greenland between 20 and 15 Ma, and then in the middle latitudes of the northern hemisphere around 2.5 Ma. Several key events have been implicated in this last great

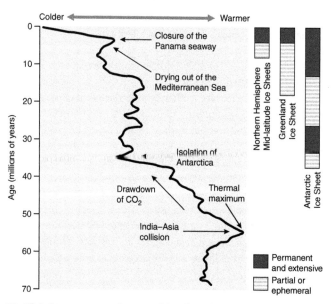

27. **Global temperature change and ice sheet development over the last 70 million years**

cooling in Earth history (Figure 27) and these are briefly reviewed next.

The uplift of the Himalayas was underway by 50 Ma after the initial collision between India and Asia. The high rainfall and steep relief in this region created a global hot spot for both physical and chemical weathering. The big rivers that drain the Himalayas transport very high loads of fine sediment and material dissolved in solution. As the rocks of the Himalayas are broken down and transported to the sea, the CO_2 from rainwater combines with the weathering products to form new compounds such as the calcium carbonate ($CaCO_3$) shells of marine creatures. As uplift continued, more and more carbon from the atmosphere became sequestered in the fossils of creatures deposited on the ocean floor. It took a very long period of weathering (about 20 million years) to sufficiently weaken the Earth's greenhouse effect to bring down mean global temperature so that large-scale glaciation could commence in the high latitudes.

As CO_2 levels continued to fall, shifting continents in the southern hemisphere resulted in the isolation of Antarctica. This was a crucial step in the initiation of large-scale glaciation in the Polar South. The opening of marine gateways between Antarctica and South America and between Antarctica and Australia produced the powerful Antarctic Circumpolar Current. This prevented warm ocean currents from reaching the waters around Antarctica and allowed the first permanent ice to accumulate about 35 Ma (Figure 27). The history of Antarctica's ice sheets is chronicled in the IRD record of the Southern Ocean. Recent work on the IRD record in marine sediments around Greenland by German geologist Jörn Thiede and his team has indicated that large-scale glaciation on that landmass was initiated much earlier than previously thought (Figure 27).

At the end of the Miocene, between 5.5 and 5 Ma, the Mediterranean Sea dried out and refilled many times as its connection with the Atlantic was disrupted by tectonic movements at the Straits of Gibraltar. The Mediterranean became an enormous salt pan as the basin's waters evaporated during times of isolation. These salt deposits can exceed 2 km in thickness. The end result was a *c*.6 per cent fall in the salt load of the global ocean so that sea water became easier to freeze. It has been argued that the extent and thickness of sea ice increased in high latitudes after this period. The enhanced albedo effect that this created would have augmented the long-term trend of Late Cenozoic global cooling.

Finally, in the Late Pliocene, sometime after 4 Ma, the formation of the Isthmus of Panama joined the continents of North and South America and closed the connection between the Atlantic and Pacific Oceans. This strengthened the Gulf Stream and increased the supply of moisture to the land masses around the North Atlantic. The final closure preceded the earliest evidence for large-scale glaciation in the northern hemisphere south of 60°N. The proximity of these events makes this an attractive hypothesis (Figure 27) for the initiation of ice sheet growth in the high middle latitudes of North America and Europe although some computer models suggest that the warming influence of the Gulf Stream may have had the opposite effect.

Over the past 55 million years, a succession of processes driven by tectonics combined to cool our planet. It is difficult to isolate their individual contributions or to be sure about the details of cause and effect over this long period, especially when there are uncertainties in dating and when one considers the complexity of the climate system with its web of internal feedbacks. The tipping point for the initiation of ice sheet formation in the high middle latitudes of the northern hemisphere may have been albedo effects and an orbital nudge that intensified the long-term cooling trend driven by tectonic processes. Once set in glacial mode, the length of glacials and interglacials was paced by the Milankovitch Cycles.

The ice age transformed

The study of marine sediments utterly transformed ice age research in ways that no one could have imagined. The discoveries of the 1970s were so profound and their implications so far-reaching that at least one leading Professor of Quaternary Science, David Bowen, ranked them in significance alongside the plate tectonics revolution. The oxygen isotope record provided a powerful proxy for long-term changes in *global* ice volume. The marine record of IRD allowed the geography of ice sheet development and decay to be reconstructed in both hemispheres with much greater spatial resolution.

It was then established in the late 1980s that catastrophic releases of icebergs from calving bays were responsible for the deposition of distinctive Heinrich layers in the marine record across the North Atlantic. The floating ice that deposited them was also shown to be a powerful agent of climate change. It became clear that the glacial stages were themselves punctuated by abrupt climate shifts. This kind of ice sheet behaviour could not be detected in the first isotope records from the tropical oceans. Rapid climate change triggered by ice sheet–ocean–atmosphere interactions was now firmly on the agenda. At about the same time that Heinrich Events were discovered in the North Atlantic, new data from the study of Greenland ice cores became available. In a remarkable twist, the ice sheets themselves would soon provide the most finely resolved records of glacial stage climate and shed new light on the causes of ice age climate change.

Chapter 9
Ice cores, abrupt climate shifts, and ecosystem change

> …ice contains no future, just the past, sealed away…clear and distinct.

> Haruki Murakami (1995)

In 1958, at the height of the Cold War, the USA developed a secret plan for a network of mobile nuclear missile launch sites beneath the Greenland ice sheet. It was codenamed Project Iceworm. Feasibility studies, involving experimental ice drilling and tunnelling, took place at Camp Century (77°N) in the northwest corner of Greenland. Sub-glacial living quarters were constructed for 250 personnel. It soon became apparent, however, that the motion of the ice was crushing the tunnels. The project was abandoned in 1966: but not before drilling trials retrieved the first long ice cores from a polar ice sheet. By drilling ice cores and melting the layers of ice for analysis, it is possible to obtain remarkably detailed records of the ice age past.

Willi Dansgaard (1922–2011) (Figure 28) was the first scientist to demonstrate that the ice sheets themselves provided an extended record of Earth's climate history. In the early 1960s, when Nick Shackleton was perfecting the measurement of oxygen isotopes on benthic foram shells in Cambridge, Dansgaard was setting up his own isotope laboratory in Copenhagen, but with very different kinds of samples. Dansgaard was interested in oxygen isotope

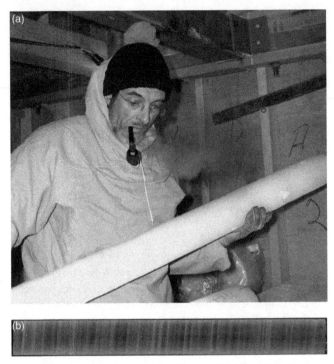

28. (a) Willi Dansgaard in Greenland in 1979 (b) Annual banding in
a 1m section of ice core from Greenland from a depth of 1837 to
1838 metres

ratios in rainfall, snow, and ice. He made the landmark discovery
that the oxygen isotope profile in ice cores provided a long-term
record of changing air temperature in the Polar regions. He was
able to show that as air temperature falls, more molecules of H_2O
containing the heavy (oxygen-18) isotope condense and are lost
from clouds as rain and snowfall. Thus atmospheric water vapour
becomes more and more depleted of ^{18}O in a poleward direction.
Dansgaard analysed rainfall samples and temperature data from
around the world to test this idea—he even collected samples in
beer bottles in his back garden in Copenhagen.

Rapid climate change

In 1966, the Americans obtained a 1,390 m ice core from Camp Century—the first ice core to penetrate the Greenland ice sheet down to bedrock. Greenland ice cores typically have very clear banding (Figure 28b) that corresponds to individual years of snow accumulation. This is because the snow that falls in summer under the permanent Arctic sun differs in texture to the snow that falls in winter. The distinctive paired layers can be counted like tree rings to produce a finely resolved chronology with annual and even seasonal resolution. The section of ice core shown in Figure 28b was produced by snow that fell around 16,250 years ago. Recent work on Greenland ice cores has allowed the end of the Pleistocene epoch and the onset of the Holocene interglacial to be dated very precisely to 11,700 years before AD 2000.

By sampling each layer of ice and measuring its oxygen isotope composition, Dansgaard produced an annual record of air temperature for the last 100,000 years. He had produced the first annual weather report for the last glacial stage. Perhaps the most startling outcome of this work was the demonstration that global climate could change extremely rapidly. Dansgaard showed that dramatic shifts in mean air temperature (>10°C) had taken place in less than a decade. These findings were greeted with scepticism and there was much debate about the integrity of the Greenland record, but subsequent work from other drilling sites vindicated all of Dansgaard's findings.

Dansgaard was to ice core research what Shackleton was to the study of marine sediments. It was therefore fitting when, in 1995, the Royal Swedish Academy of Sciences awarded its prestigious Crafoord Prize (widely regarded as the equivalent of a Nobel Prize) jointly to Shackleton and Dansgaard for their work on isotopes in geoscience.

The ice age atmosphere

As layers of snow become compacted into ice, air bubbles recording the composition of the atmosphere are sealed in discrete layers. This fossil air can be recovered to establish the changing concentration of greenhouse gases such as carbon dioxide (CO_2) and methane (CH_4). The ice core record therefore allows climate scientists to explore the processes involved in climate variability over very long timescales.

A Swiss physicist, Hans Oeschger (1927–98), made fundamental contributions to our understanding of ice age climate change. He pioneered the measurement of greenhouse gases in the bubbles trapped in ancient ice. In his laboratory at the University of Bern, Oeschger analysed many thousands of samples from Greenland and Antarctica. In 1979 his team was the first to show that CO_2 concentrations during glacial stages were almost half those of the present. Oeschger showed that atmospheric CO_2 concentrations were about 180 ppm (parts per million) during glacials, but around 280 ppm during interglacials.

Ice accumulation is generally much slower in Antarctica, so the ice core record takes us much further back in time. The lower part of Figure 29 shows an 800,000-year Antarctic ice core record that spans eight glacial–interglacial cycles. This was produced by the European Project for Ice Coring in Antarctica (EPICA). Note how the changes in temperature closely track the changes in methane and CO_2. Methane is a potent greenhouse gas—it is stored in large volumes in the frozen biomass of the permafrost and as methane hydrate within sediments beneath the ocean floor. Ice core data have been fundamental in demonstrating that changes in the composition of the atmosphere played a key role in the shifting climates of the Quaternary, but there is still much debate about the processes involved in the leads and lags. CO_2 exchange between the oceans and atmosphere is the key link between the Milankovitch Cycles and the glacial and interglacial shifts of the

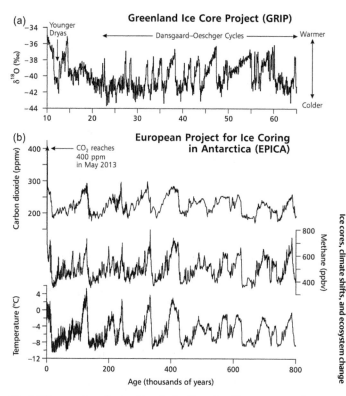

29. (a) Temperature fluctuations in the Greenland ice core record between 65,000 and 10,000 years ago (b) An 800,000-year record of CO_2, methane, and air temperature from Antarctica. Note the different timescales for the Greenland and Antarctic records

Quaternary. A key challenge, however, is to develop a better understanding of the processes that lead to marked shifts in atmospheric greenhouse gases during the course of glacial and interglacial cycles.

The ice age does not give up its secrets easily. Ice coring involves huge logistical challenges—pushing equipment and people to the limit. The Icelandic geophysicist Sigfús Johnsen (1940–2013) designed the specialist ice drilling equipment for the later Greenland projects. Bone chilling temperatures and powerful winds make the elevated interiors (where the ice is thickest) of the polar ice sheets some of the most inhospitable places on Earth. The lowest temperature ever measured at the Earth's surface (−89.2°C) was recorded at the Vostok Station in the middle of the East Antarctic Ice Sheet on 21st July 1983. Vostok also holds the record for the longest ice core. In late 2012, a Russian team drilled 3,768 m to reach Lake Vostok: the largest of the sub-glacial lakes in Antarctica and the third largest lake by volume in the world.

Dansgaard–Oeschger Events

The ice core records from Greenland reveal a remarkable sequence of abrupt warming and cooling cycles *within* the last glacial stage. These are known as Dansgaard–Oeschger (D–O) cycles. The upper part of Figure 29 shows a series of D–O cycles between 65,000 and 10,000 years ago when mean annual air temperatures on the Greenland ice sheet shifted by as much as 10°C. Twenty-five of these rapid warming events have been identified during the last glacial period. This discovery dispelled the long held notion that glacials were lengthy periods of stable and unremitting cold climate. The ice core record shows very clearly that even the glacial climate flipped back and forth.

There has been much discussion about the cause of these D–O cycles and their relationship to the Heinrich Events. Changes in ocean circulation triggered by influxes of freshwater from the continents have been implicated. D–O cycles commence with a very rapid warming (between 5 and 10°C) over Greenland followed by a steady cooling (Figure 29). The most pronounced coolings are associated with Heinrich Events when severe glacial climates with extreme cold and aridity characterized many parts of the northern hemisphere.

The great ocean conveyor

Figure 30 shows the circulation system linking the world's oceans that moves vast quantities of heat around the globe. This circulation includes warm, shallow currents and deeper flows of denser, colder water. It is known as the thermohaline circulation because it is driven by contrasts in water temperature (thermo) and salinity (haline). These properties control the density of seawater. It takes about 1,000 years for a body of water to complete a full cycle. This circulation has been called the great ocean conveyor—it plays a fundamental role regulating global climate.

There are strong thermal gradients in both hemispheres because the low latitudes receive the most solar energy and the poles the least. To redress these imbalances the atmosphere and oceans move heat polewards—this is the basis of the climate system. In the North Atlantic a powerful surface current takes warmth from the tropics to higher latitudes: this is the famous Gulf Stream and its northeastern extension the North Atlantic Drift. Two main forces drive this current: the strong southwesterly winds and the return flow of colder, saltier water known as North Atlantic Deep

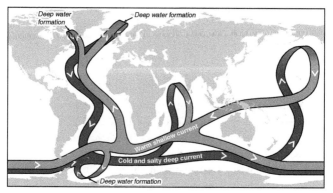

30. Ocean currents: the thermohaline circulation

Water (NADW). The surface current loses much of its heat to air masses that give maritime Europe a moist, temperate climate. Evaporative cooling also increases its salinity so that it begins to sink. As the dense and cold water sinks to the deep ocean to form NADW, it exerts a strong pull on the surface currents to maintain the cycle. It returns south at depths >2,000 m.

The Younger Dryas cooling

The thermohaline circulation in the North Atlantic was periodically interrupted during Heinrich Events when vast discharges of melting icebergs cooled the ocean surface and reduced its salinity. This shut down the formation of NADW and suppressed the Gulf Stream. The warming at the time of the last deglaciation also stalled abruptly around 12,800 years ago bringing Arctic conditions back to much of the northern hemisphere for about 1,200 years. This rapid cooling is known as the Younger Dryas after the arctic/alpine evergreen dwarf shrub (*Dryas octopetala*) whose pollen is common in European vegetation records from this period. The mean annual temperature at the summit of the Greenland ice sheet may have been about 15°C colder than present during the Younger Dryas. This cold snap lasted long enough to allow glaciers to advance in many parts of the world. Glaciers reformed in the uplands of Scotland, England, and Wales. Arctic birds, insects, mammals, and fish headed south once again.

Cause of the Younger Dryas cold snap

The Younger Dryas cooling may have been triggered by huge volumes of freshwater flowing into the high latitude North Atlantic from glacier-lake outburst floods. One of the largest glacial lakes formed by the melting Laurentide ice sheet is known as Lake Agassiz. When it emptied catastrophically around 12,800 years ago it created a freshwater cap on the ocean surface that halted the formation of NADW and the Gulf Stream stalled. This cooling event shows up very clearly in the GRIP ice core record (Figure 29). Many of the ideas about the linkages between ocean circulation and abrupt climate change have been shaped by Wally Broecker of Columbia University's Lamont-Doherty Earth Observatory.

Box 14
Glacial Lake Agassiz

The existence of this great lake was first postulated in 1824 by the American geologist William Keating (1799–1844). It was named Lake Agassiz in 1879, six years after the death of the celebrated Harvard professor who had become the most famous scientist in America. At its maximum extent it was the largest lake on the North American continent (>500,000 km²) with a volume greater than all of the modern Great Lakes combined. It extended across parts of Manitoba, Saskatchewan, and Ontario in Canada, and North Dakota and Minnesota in the USA. Fed by the wasting Laurentide ice sheet, it rose and fell repeatedly during its 4,500-year existence at the end of the last glacial stage. At times, the present location of Winnipeg was submerged beneath more than 200 m of water. From time to time, rapid outflows produced catastrophic floods and there is currently a lively debate about whether the floodwaters that triggered the Younger Dryas cooling flowed northwards down the Mackenzie River into the Arctic Ocean or directly into the North Atlantic via the St Lawrence.

J Harlen Bretz (1882–1981)

J Harlen Bretz was a geologist who discovered evidence for truly catastrophic floods that took place at the end of the last glacial stage on the Columbia River Plateau in eastern Washington State. Bretz described large-scale erosion features including the enormous basalt canyons of the Columbia River gorge that he named the Channeled Scablands. He published his great flood theory in 1923 exactly 100 years after William Buckland's

Reliquiae Diluvianae. Perhaps this coincidence was not lost on members of the American geological establishment, who were vehemently opposed to what they saw as a new deluge theory. It was Joseph Pardee (1871–1960) of the United States Geological Survey who suggested that the floodwaters originated from Glacial Lake Missoula in western Montana where a southern lobe of the Cordilleran ice sheet created an ice dam over 25 km wide and more than 600 m high. This dam periodically failed, releasing about 2,100 km^3 of raging floodwater—the equivalent of half the volume of today's Lake Michigan. Time and again the Missoula floods swept through the Scablands and on to the Pacific Ocean as the ice dam reformed and failed. Recent research has suggested that there may have been about 40 of these outburst floods at the end of the last glacial stage between c.15,000 and 13,000 years ago. Bretz has been hailed as a pioneer of neocatastrophism: the movement that recognized the importance of very high magnitude events in landscape evolution. His pioneering work on palaeofloods inspired a much later generation to consider catastrophic meltwater discharges as a potential cause of abrupt climate change.

The ice core records heralded a new era in climate science: the study of abrupt climate change. Most sedimentary records of ice age climate change yield relatively low resolution information—a thousand years may be packed into a few centimetres of marine or lake sediment. In contrast, ice cores cover *every* year. They also retain a greater variety of information about the ice age past than any other archive. We can even detect layers of volcanic ash in the ice and pinpoint the date of ancient eruptions. New questions have emerged: how did ecosystems and humans respond to these abrupt changes? Were they regional or global phenomena? What role did they play in the demise of the ice age megafauna?

Late Pleistocene ecosystems and extinctions

Figure 31 illustrates the major ecosystem reorganizations that took place across Europe during the rapid environmental changes of the last glacial to interglacial transition. The maps show the presence or absence of deciduous oak from a large number of vegetation records obtained from sediment cores collected from lakes and peat bogs across the continent. Fossil pollen has been extracted from these cores and counted, layer by layer, to build up a picture of long-term

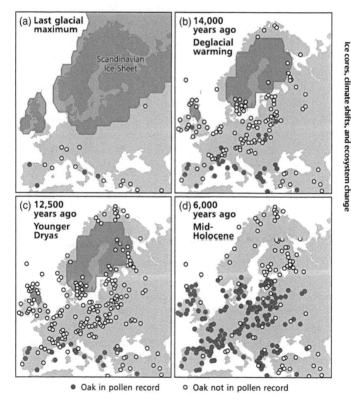

● Oak in pollen record   ○ Oak not in pollen record

31. The dispersal of oak across Europe since the Last Glacial Maximum

vegetation change. Four time slices are given: from the Last Glacial Maximum (LGM) to the middle of the present Holocene interglacial. The open circles in central and northern Europe indicate the presence of pioneer species such as birch, juniper, and pine as well as open ground or steppe plants like grasses and sedges. The records have been dated using radiocarbon.

At the LGM the Arctic tundra that fringed the ice sheets was dominated by a landscape of permafrost, frost-shattered rocks, thin soils, and a low vegetation adapted to brief, cold, growing seasons. The upper layer of permafrost thawed in the glacial summer creating shallow wetlands and boggy soils where grasses, mosses, herbs, sedges, and other flowering plants briefly flourished. These shallow-rooting plants clustered together to conserve warmth and resist attack from strong winds and blowing snow. Arctic willow (*Salix arctica*), the northernmost woody plant on Earth, is found in central European pollen records from the last glacial stage. It was an important food source for muskox, mammoth, bison, and arctic hare. Arctic plant communities and permafrost returned to many parts of central Europe during the Younger Dryas cooling. Relics of this flora are found today in the far north and in high mountains.

In the Mediterranean, far to the south of the erosive power of the ice sheets, several lakes are very ancient indeed, with unbroken sedimentary records than span multiple glacial–interglacial cycles. These very long records of vegetation change show Milankovitch cyclicity as well as clear responses to Heinrich and Dansgaard–Oeschger events during the last glacial stage. Deciduous oak is a key indicator of warm and moist interglacial conditions. It is absent from central and northern Europe during glacial stages. At the LGM it is only recorded at sites in the Mediterranean (Figure 31a). Southern Europe contains zones of high biodiversity where species of plants and animals have survived throughout the Quaternary ice age. These zones are called *refugia*: they are crucial for the long-term survival of Europe's forest ecosystems. Oak, along with other temperate interglacial tree species such as beech, elm, and alder (Figure 4), survived the glacial stages at intermediate elevations in the wetter parts of the Mediterranean mountains.

The lakes and peat bogs of northern Europe formed after the recession of the last ice sheets, so their sediment records are much shorter than those in the south. By 14,000 years ago, a large number of vegetation records are available for central and northern Europe (Figure 31b) because meltwater streams and thawing permafrost produced extensive wetlands where pollen and plant macrofossils were readily preserved. The vegetation records from these settings track the return of pioneer plants as the ice sheets receded. The last deglaciation was well underway by 14,000 years ago and the ice sheets were much reduced in size (Figure 31b). Under a rapidly warming climate, oak had spread northwards from its southern refugia as far as the latitude of northern Germany. By 12,500 years ago, however, in the middle of the Younger Dryas cooling, this expansion had halted—oak is only found south of the Alps at this time (Figure 31c). By the middle of the Holocene, some 6,000 years later, oak had expanded its range as far as southern Scandinavia and a full interglacial flora with mature soils was well established across temperate Europe.

The return of the forests to central and northern Europe at the end of glacial stages is dependent upon migration from southern refugia. If trees fail to survive in the Mediterranean they become extinct. The marine record tells us that glacial periods last for about 80,000 years and interglacials about 15,000 years. For most of the Quaternary deciduous forests have been absent from most of Europe. The D–O warmings were too brief for significant forest expansion to take place. All this means that the interglacial forests of temperate Europe that are so familiar to us today are, in fact, rather atypical when we consider the long view of Quaternary time. Furthermore, if the last glacial period is representative of earlier ones, for much of the Quaternary terrestrial ecosystems were continuously adjusting to a shifting climate.

Box 15
Pleistocene megafauna

Figure 31b shows how rapidly the vast tundra steppe biome contracted as it was colonized by pioneer woodland species (open circles) at the end of the last glacial period. The close of a glacial was always a stressful time for mammoths and other large mammals in northerly latitudes because their habitats shrank and populations became increasingly fragmented and isolated. The demise of the Pleistocene megafauna (including the mammoth, Irish elk, woolly rhino, and others) is a hotly disputed topic but their extinction cannot be attributed to a single cause. Contrary to earlier ideas that the megafauna died out at roughly the same time at the end of the last glacial, we now know that some of these animals persisted in isolated refugia for several thousand years into the present Holocene interglacial. Populations of mammoth, for example, survived on Wrangel Island in the East Siberian Sea until about 4,000 years ago. They outlived the last mammoths on the Siberian mainland by about 6,000 years. The Irish elk (*Megaloceros giganteus*) persisted in western Siberia until about 7,700 years ago.

Such extinctions are now viewed in terms of prolonged periods of decline rather than abrupt terminations of species by climate change or hunting. The polar bear may well be the next one to go. The latest research is attempting to identify the precise timing and geography of last glacial and Holocene extinction patterns for each species by compiling large databases of radiocarbon-dated fossils. This information is being integrated with genetic data and with high resolution records of vegetation and climate change to better understand the ecology of extinction. A role for human hunters sending isolated populations over the edge of viability cannot be discounted. Indeed, this ice age menagerie had successfully negotiated previous glacial to interglacial transitions. The last one, however, was very different in one key respect: modern humans were present across Europe, Siberia, and North America for the first time.

Humans in the last glacial

The Neanderthals were another casualty of the last glacial period and numerous theories have been advanced to account for their demise. Pinning down the timing of their final disappearance is problematic because many Neanderthal fossils lie close to the limit of radiocarbon dating and this can increase the chances of sample contamination from more recent radiocarbon. Anatomically modern humans (*Homo sapiens*) spread rapidly across Europe between 50,000 and 35,000 years ago from their original homeland in Africa. This placed them in direct competition with Neanderthal groups. The latest research indicates that Neanderthals did not survive much beyond 35,000 years ago. It has been suggested that this new competition and the very severe climatic conditions of Heinrich Event 4 (just after 40,000 years ago; Figure 26c) may have been enough to sink the final nails into the Neanderthal coffin. There are clear parallels here with the megafaunal extinctions: we need more reliable dates and a better understanding of the geography and ecology of the Neanderthal demise.

Palaeolithic archaeologist Clive Gamble and colleagues at the University of Southampton have used >2,000 radiocarbon dates from archaeological sites across western Europe as a proxy for ice age population change after the Neanderthal extinction. They have explored population dynamics for modern humans in relation to the Greenland ice core temperature record for the period from 30,000 to 6,000 years ago (Figure 32). This analysis indicates that the dispersal of modern humans across Europe took place within a wide range of climatic tolerances. Whilst southern Europe was an important refuge around the time of the LGM, it also points to tolerance of extreme cold in the northern regions. The number of archaeological sites with reliably dated evidence of human presence increases rapidly in France and northern Europe between 16,000 and 14,000 years ago. The Younger Dryas event

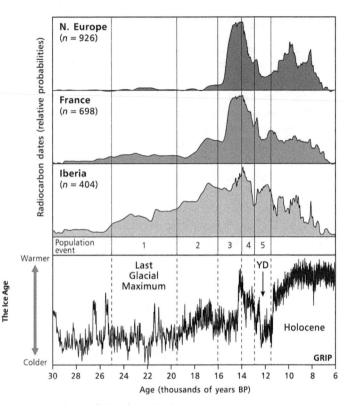

32. Ice age population dynamics in Europe

sees a decline in these regions at this time but an increase in Iberia. This novel approach to the study of humans in the ice age is based on several assumptions and there are issues surrounding the differential preservation of archaeological sites, but it does allow us to view the big picture of social change and human–environment interaction at a continental scale.

Modern humans were able to develop very effective strategies to seek out resources and cope with extreme cold in the barren

33. Mammoth hunters in the last glacial period

tundra biomes of ice age Europe, Asia, and North America. Heinrich Events would have posed especially harsh challenges in the northern regions. Figure 33 shows a group of mammoth hunters—it is a faithful reconstruction by Giovanni Caselli based on archaeological data from the Upper Palaeolithic site of Dolní Věstonice in the Czech Republic. This wonderful image captures the social complexity, sophisticated division of labour, and inventive use of resources that was needed for humans to thrive in the rapidly shifting environments of the last glacial period.

Epilogue

The history of ice age research reveals a good deal about the history of geology and the changes that have taken place in the way we think about the Earth. Lyell's uniformitarian straitjacket was loosened in the second half of the 20th century so that most geologists now take the view that Earth history involved extended periods of gradual change punctuated by infrequent catastrophic events. The last glacial stage was an especially eventful period—great floods and floating ice are very much back on the agenda.

The modern era of debating climate change can be traced back to the discovery of the Adams Mammoth in 1799. But the discourse of the early and mid-19th century was dominated by those who denied the existence of climate change in the recent geological past. The context, of course, was very different from today but powerful individuals with vested interests made their voices heard. The case for the glacial theory and ice age climate change succeeded because its advocates presented a huge body of irrefutable evidence. Winning this argument took the best part of 60 years.

A view then emerged in the last century that glacials were few in number and Quaternary climate change was a slow process paced over many thousands of years. Even for the glacial–interglacial cycles of the Quaternary, there was nothing to suggest that anything other than a sensible uniformitarian notion of gradual change was needed. These ideas were crushed by the marine and ice core records. The traditional view of grand climate rhythms with smooth, orderly transitions from warm to cold and back again was no longer tenable.

If big bones and boulder clay were the staples of 19th-century ice age research, microfossils and isotopes came to dominate the next, and the scale of analysis went global. The Milankovitch revival and the development of scientific methods of dating stand out as key advances. Modern Quaternary science involves many disciplines and a bewildering array of approaches to retrieve information about the past. The woolly mammoth is now the first extinct mammal to have its DNA sequence decoded. Hair from the Adams Mammoth has been used to retrieve a complete mitochondrial genome along with DNA sequences from the nuclear genome. There is currently much excited speculation about cloning the mammoth but much less critical reflection on the whereabouts of a viable long-term habitat for this ice age beast.

The Quaternary is a unique natural laboratory with a remarkable variety of exceptionally well preserved records. This invites us to ask fundamental questions about the natural world and how its components interact at global, regional, and local scales. These records also provide a meeting place for geology and archaeology. Key chapters in the story of human evolution and dispersal took place during the shifting environments of the Quaternary ice age. We now know that ice age ecosystems were rapidly and repeatedly transformed as plants, animals, and humans reorganized their worlds.

The ice cores yielded stunning datasets that forced all disciplines to rethink their understanding of ice age change. A new tier of climate variability was revealed. The Heinrich layers in the North

Atlantic demonstrated the connections between ice sheets, oceans, and atmosphere: icebergs could change the climate. It became clear that the climate system is finely tuned and prone to abrupt change. A new agenda was forged—not only for all aspects of ice age research, but for global climate science. In many respects the scientific community and its climate modellers are still coming to terms with the reality of rapid climate change. The search for the underlying causes and key feedback processes is still very much in progress.

Hans Oeschger was one of the first scientists to warn of the dangers of an enhanced greenhouse effect driven by anthropogenic increases in carbon dioxide (CO_2). It would be impossible to overstate how deeply this work has penetrated the mainstream of present-day debate. As one of the lead authors of the First Assessment Report of the Intergovernmental Panel on Climate Change (IPCC), Oeschger made the bold leap from ice core science to global climate policy. Because continuous measurements of atmospheric CO_2 from the Mauna Loa Observatory in Hawaii only go back to 1958, it was the ice core records for the last few centuries that first showed the striking rise in greenhouse gases following the Industrial Revolution. The ice core data allowed climate scientists to place recent trends in long-term context. These trends are, quite literally, off the Quaternary scale. The baseline CO_2 value for interglacials is *c.*280 parts per million (ppm). On 9th May 2013 the concentration of atmospheric CO_2 exceeded 400 ppm for the first time since the balmy conditions of the Pliocene when sea level was more than 20 m higher than today.

We still need to better understand the processes involved in rapid climate warming—especially the interactions between ice, oceans, and atmosphere. A deeper understanding of the interactions between landscapes, ecosystems, and shifting climates is also needed. Paradoxically, in an era of warming climate, the study of the ice age past is now more important than ever.

Publisher's acknowledgements

Epigraph

Archibald Geikie (1887)
Archibald Geikie, The Scenery of Scotland, London, Macmillan and
   Co., 2nd edition, (1887)

Chapter 1: The Quaternary ice age

Nothing excites the imagination more than the study of the
   Quaternary.
Maurice Gignoux (1955)
Stratigraphic Geology, W.H. Freeman, San Francisco, 682 pages
   (1955)

Chapter 2: Erratic boulders and the diluvium

Time it does not matter
But time is all we have
To think about
Deep Purple (2013) (Airey, Gillan, Glover, Morse, Paice)
A Simple Song from the Deep Purple album *Now What?!* (2013)

Chapter 3: Monster glaciers

…it seemed as if Nature was stepping out of its normal course, and
   the glaciers expanded rapidly…
Bernhard Friedrich Kuhn (1787)

Versuch über den Mechanismus der Gletscher (Investigation into the
mechanics of glaciers), *Magazin für die Naturkunde Helvetiens*,
Volume 1, pp. 119–36 (1787)

Chapter 4: *Die Eiszeit*

Imagination will often carry us to worlds that never were, but without
it we go nowhere.
Carl Sagan (1980)
Carl Sagan, *Cosmos*. New York: Random House, 1st edn,
(1980)

Chapter 5: 1840

Ideas without precedent are generally looked upon with disfavour
and men are shocked if their conceptions of an orderly world
are challenged.
J Harlen Bretz (1928)
Engraved on a plaque honouring Bretz outside the Dry Falls Museum
in Coulee City, Washington State

Chapter 6: Ice sheets or icebergs?

Concerning the Glacial period, geologists hold the most varied
opinions, both with regard to its origin and to the mode of
action of the ice.
Thomas Belt (1877)
Man and the glacial period, *Popular Science Monthly*, 12, pp. 61–74
(1877)

Chapter 7: Glacials, interglacials, and celestial cycles

The imperfection of all our records of the past is too well known to
geologists.
Alfred Russel Wallace (1879)
Glacial epochs and warm polar climates, *Quarterly Review*,
pp. 119–34 (1879)

Chapter 8: Deep ocean sediments and dating the past

The sea must know more than any of us.
Carl Sandburg (1918)
The Sea Hold by Carl Sandburg from *Cornhuskers*. New York: Henry
    Holt and Company, (1918)

Chapter 9: Ice cores, abrupt climate shifts, and ecosystem change

... ice contains no future, just the past, sealed away ... clear and
    distinct.
Haruki Murakami, *Blind Willow, Sleeping Woman* (1995)
Published in English by Harvill Secker, 1st edn (2006)

Publisher's acknowledgements

References

In addition to the works listed in the Further reading section and given in the main text, the following sources were especially useful as I researched this book. If you want to read more about early geology and the history of the glacial theory, the magnificent volumes by Martin Rudwick are strongly recommended.

E.C. Agassiz, *Louis Agassiz, His Life and Correspondence* (Volume 1), Houghton, Mifflin and Company (1885)

D.J. Blundell and A.C. Scott (eds), *Lyell: the Past is the Key to the Present*, Geological Society Special Publication, 143: 376 (1998)

D.Q. Bowen, *Quaternary Geology*, Pergamon (1978)

J. Bowlby, *Charles Darwin: A New Biography,* Pimlico (1991)

W. Boyd Dawkins, *Cave Hunting: Researches on the Evidence of Caves Respecting the Early Inhabitants of Europe*, Macmillan and Co. (1874)

P.J. Boylan, 'The role of William Buckland (1784–1856) in the recognition of glaciation in Great Britain'. In: J. Neale and J.R. Flenley (eds), *The Quaternary in Britain*, 1–8, Pergamon Press (1981)

T.P. Burt, R.J. Chorley, D. Brunsden, N.J. Cox, and A.S. Goudie, *The History of the Study of Landforms or The Development of Geomorphology*, Volume 4: *Quaternary and Recent Processes and Forms (1890–1965) and the Mid-Century Revolutions*, The Geological Society (2008)

R.J. Chorley, A.J. Dunn, and R.P. Beckinsale, *The History of the Study of Landforms or The Development of Geomorphology*, Volume 1: *Geomorphology Before Davis*, Methuen (1964)

E.S. Dana et al., *A Century of Science in America*, Yale University Press (1918)

Charles Darwin's letter to William Boyd Dawkins, reproduced in this book, comes from: <http://darwin-online.org.uk/> (1873)

J.M. Geikie, *The Great Ice Age and its Relation to the Antiquity of Man*, D. Appleton and company (1874)

E. Gordon, *The Life and Correspondence of William Buckland*, John Murray (1894)

J.M. Grove, *The Little Ice Age*, Methuen (1988)

G.L. Herries Davies, 'The Tour of the British Isles Made by Louis Agassiz in 1840', *Annals of Science*, 24: 131–46 (1968)

G.L. Herries Davies, *North from the Hook: 150 Years of the Geological Survey of Ireland*, Geological Survey of Ireland (1995)

R. Huxley (ed.), *The Great Naturalists*, Thames and Hudson (2007)

A. O'Connor, 'The Competition for the Woodwardian Chair of Geology: Cambridge, 1873', *British Journal for the History of Science*, 38: 437–62 (2005)

W.F. Ruddiman, *Earth's Climate: Past and Future*, W.H. Freeman, 2nd edn (2007)

M.J.S. Rudwick, *Worlds before Adam: The Reconstruction of Geohistory in the Age of Reform*, University of Chicago Press (2008)

M.J.S. Rudwick, *Bursting the Limits of Time: The Reconstruction of Geohistory in the Age of Revolution*, University of Chicago Press (2005)

J.C. Woodward, 'Quaternary Geography and the Human Past'. In: N. Castree, D. Demeritt, D.M. Liverman, and B.L. Rhoads (eds), *A Companion to Environmental Geography*, Wiley-Blackwell, pp. 298–322 (2009)

H.B. Woodward, *The History of the Geological Society of London*, Geological Society (1907)

H.B. Woodward, *History of Geology*, Watts and Co. (1911)

W. B. Wright, *The Quaternary Ice Age*, Macmillan and Co. (1914)

Some useful websites

Keith Montgomery at the University of Wisconsin (Marathon County) maintains an excellent series of web resources entitled: 'Debating Glacial Theory 1800–1870':
<http://www1.umn.edu/ships/glaciers/

Earthguide is part of the Geosciences Research Division at Scripps Institution of Oceanography in California. The following website contains links to a rich assortment of ice age materials: <http://earthguide.ucsd.edu/>

In 2008, the Linda Hall Library in Kansas City, Missouri, held an exhibition of rare books and journals entitled 'Ice: A Victorian Romance'. It documented the 19th-century exploration of the polar world and the origins of the glacial theory: <http://www.lindahall.org/events_exhib/exhibit/exhibits/ice/index.shtml>

Portraits of key figures in the development of British geology can be found in *The Oxford Dictionary of National Biography*: <http://www.oxforddnb.com/>

Most researchers who delve into the Quaternary ice age are affiliated to the International Union for Quaternary Research (INQUA): <http://www.inqua.org/>

Further reading

Ice age environments and global change

D.E. Anderson, A.S. Goudie, and A.G. Parker, *Global Environments through the Quaternary*, Oxford University Press, 2nd edn (2013)

B. Fagan (ed.) *The Complete Ice Age: How Climate Change Shaped the World*, Thames and Hudson (2009)

A.M. Lister and P.G. Bahn, *Mammoths: Giants of the Ice Age*, University of California Press (2009)

R.B. Alley, *The Two-Mile Time Machine: Ice Cores, Abrupt Climate Change, and Our Future*, Princeton University Press (2000)

P.S. Martin, *Twilight of the Mammoths: Ice Age Extinctions and the Rewilding of America*, University of California Press (2005)

Glacial environments

M.J. Hambrey and J. Alean, *Glaciers*, Cambridge University Press, 2nd edn (2004)

D. Benn and D.J.A. Evans, *Glaciers and Glaciation*, Hodder Education, 2nd edn (2010)

I have tried to avoid excessive overlap with existing *Very Short Introductions*. The following VSIs are valuable companions to this work:

M.J. Benton, *The History of Life: A Very Short Introduction*, Oxford University Press (2008)

C. Gosden, *Prehistory: A Very Short Introduction*, Oxford University Press (2003)

M.A. Maslin, *Climate: A Very Short Introduction*, Oxford University Press (2013)

M. Redfern, *The Earth: A Very Short Introduction*, Oxford University Press (2003)

K.S. Thomson, *Fossils: A Very Short Introduction*, Oxford University Press, (2005)

B. Wood, *Human Evolution: A Very Short Introduction*, Oxford University Press (2005)

Index

Index

SOCIAL MEDIA
Very Short Introduction

Join our community

www.oup.com/vsi

- Join us online at the official Very Short Introductions
  Facebook page.
- Access the thoughts and musings of our authors with our
  online **blog**.
- Sign up for our monthly **e-newsletter** to receive information
  on all new titles publishing that month.
- Browse the full range of Very Short Introductions online.
- Read **extracts** from the Introductions for free.
- Visit our library of **Reading Guides**. These guides, written by our
  expert authors will help you to question again, why you think
  what you think.
- If you are a teacher or lecturer you can order inspection
  copies quickly and simply via our website.

Visit the Very Short Introductions website to access all this and
more for free.
www.oup.com/vsi

ONLINE CATALOGUE
A Very Short Introduction

Our online catalogue is designed to make it easy to find your ideal Very Short Introduction. View the entire collection by subject area, watch author videos, read sample chapters, and download reading guides.

http://fds.oup.com/www.oup.co.uk/general/vsi/index.html